COUNTING

COUNTING

Humans, History and the Infinite Lives of Numbers

Benjamin Wardhaugh

WILLIAM COLLINS

William Collins
An imprint of HarperCollins*Publishers*
1 London Bridge Street
London SE1 9GF

WilliamCollinsBooks.com

HarperCollins*Publishers*
Macken House,
39/40 Mayor Street Upper,
Dublin 1, DOI C9w8, Ireland

First published in Great Britain in 2024 by William Collins

1

A catalogue record for this book is available from the British Library

ISBN 978-0-00-843646-9 (Hardback)
ISBN 978-0-00-843647-6 (Trade Paperback)

Set in Adobe Garamond Pro by Palimpsest Book Production Limited, Falkirk, Stirlingshire

Printed and bound in the UK using 100% renewable electricity at CPI Group (UK) Ltd

This book contains FSC™ certified paper and other controlled
sources to ensure responsible forest management.

For more information visit: www.harpercollins.co.uk/green

For William, Ralph and Laurence

Contents

INTRODUCTION

What is counting?

What does it mean, to count?

A woman collects shells, pierces them, threads them one by one onto a leather strip. Ties the strip, wears it.

People creep into a sacred space deep beneath the earth bearing torches, water and pigment. At specially chosen places, they mark the walls with hand signs: *one finger, two fingers, three fingers* . . .

A scribe squats on the ground in the greatest city in the world; marks a slab of clay with symbols that, for him and his people, mean *two, three, five, goats, grain sacks.*

A sophisticated, literate Athenian citizen spends his day exchanging counters, voting tokens and coins, in an elaborate dance that determines the outcome of trials, gains him food to eat, reaffirms his status in the city and the world.

A weary Dutch businesswoman pores over a table of symbols in a handwritten ledger, checking, copying and correcting until the text matches up with reality.

A young Korean girl obsessively checks and rechecks her phone to see how many likes her latest vlog post has accumulated.

A Tongan woman utters a traditional, special set of counting words as she assembles hundreds of strips of pandanus ready to make a woven mat.

A Mayan king, deep in trance, presides as a new monument is dedicated in his capital city, adorned with elaborate symbols representing number, time and the gods.

The story of counting is as wide, deep and tangled as the story of human culture. It is the story of human attempts to find some order in an unruly world; or, perhaps, to impose on a reluctant world the order that humans find within themselves. Very nearly every culture documented in history has counted in one way or another, usually in several. The huge array of different ways people count, and of reasons why they do so, reflect their different preferences and preoccupations, their ways of thinking and being.

Counting underpins a vast range of activities, from census taking and food management, to assessing your popularity or tracking appointments and anniversaries. It leaves traces in the archaeological record across tens of thousands of years, starting long before cities, agriculture or writing. It stands at the root of science and technology, and it has often been suggested that if humans make contact with species on other planets, one of the first things to talk about with them – perhaps even the subject with which to learn to talk in the first place – will be counting.

But what *is* it?

'Counting' can seem like an unruly grab-bag of almost totally unrelated actions; a label covering a huge set of very different cultural practices. The range of different activities called counting seems too wide for comfort, and at least superficially, it is not clear what they all have in common; or even whether they have anything in common at all.

Almost any definition of counting is problematic, but one of the best is attributed to the seventeenth-century German philosopher Gottfried Leibniz. It says that counting is repeated attention. Counting is what happens when you think 'this . . . this . . . this . . . this', and have some way of keeping track.

The means of keeping track might be a set of words or a

set of symbols; it might be a set of tally marks or a set of beads on a string. There are several other possibilities. But if you are paying repeated attention to objects or events and you have some way of keeping track of that process, then you are counting.

It is in the different ways of keeping track, in fact, that much of the enormous diversity of human counting takes place.

———

Counting is different from measuring, which is concerned with comparing one object with another, although symbols that record the outcome of a count have a very long history of being repurposed to record the outcome of measurements as well. Counting is different from calculation, although it turns out that nearly every method of counting has at one time or another been adapted in order to do at least simple arithmetic: to add together two counts or to take one away from another.

Counting is also different from using number words or number symbols as a handy set of labels. A phone number is not the result of anybody counting anything, and 'prisoner two-four-six-oh-one' does not necessarily stand at the end of a line – real or imaginary – of 24,601 people. Although he might.

Counting has a less clear boundary in the direction of machines that count: visitor counters at the doors of shops and museums, for instance. It seems eccentric to insist that those machines are not really 'counters', yet they break any claim that counting has to involve conscious, human attention. Perhaps not every boundary can be quite distinct, particularly at a time when counting machines are changing the world so rapidly.

Animals do not spontaneously count. They certainly pay attention to features of their environment one after another, but no wild animal has been spotted devising a way to keep track of that repeated attention. And even the brightest specimens of the most promising species struggle to use counting techniques invented by humans – words, symbols – after the first few numbers. On the other hand, animals do display some of the capabilities that underlie human counting: an ability to estimate the relative size of groups of objects, in particular. In biological terms, counting didn't come from nowhere, although it does seem to be unique to humans – at least on this planet.

————

Counting does not have one single history. There are several different ways to keep track of the things you pay attention to; they have different advantages and disadvantages that become important in different situations. Counting with words, with gestures, with symbols, using machines: each has risen and fallen and risen again at different times and places. There is no way to take the world's ways of counting and arrange them in a line, from worst to best or from most primitive to most sophisticated.

The story of counting is shaped, instead, like a tree. It has several roots, many branches, and innumerable twigs and leaves. Counting has grown and travelled with the human species, ramifying into very nearly every culture past and present. Sometimes it is possible to follow a single branch for some distance: sometimes a branch turns out to cross, to touch (or nearly touch) other branches. The number symbols that were invented in India, and now dominate the world, are like this. It is possible to follow them from their origin through their long – and ongoing – peregrination across the world, and to watch their interaction with many other traditions of counting along the way.

Elsewhere there are groups of branches – perhaps, better, the twigs within a single branch – with something definite in common. A preference for counting devices in East Asia: rods, the abacus and microchips. A preference for words and gestures in Oceania.

This book is shaped like a tree, too. First, there are two chapters about the roots of counting: the features of human cognition and anatomy, and of the Stone Age environment, that make it possible for humans to count, and that provide the most basic, pervasive and enduring ways to keep track of different objects or events. Humans have innate abilities that are relevant to counting, as well as perhaps an innate habit of spontaneously focusing on quantity. And the earliest available ways of keeping track of a count are counters, fingers, tally marks and words: technologies that will turn up again and again in the different branches of the world's story of counting.

Then there are six chapters about different branches of counting's story, organised as a world-wrapping journey that follows the great human spread out of Africa: to the Near East, Europe, South and East Asia, and on into Oceania and finally the Americas. These emphasise what is most characteristic in each part of the world: the invention and use of number symbols in the Fertile Crescent, for instance, or counters and counting boards in Europe. One chapter is about the Indian number symbols, and necessarily spans the world in pursuit of their story. Further east, the book emphasises counting machines in East Asia and counting words in the Pacific. Different choices could have been made; no part of the world has an exclusive preference in its ways of counting.

The stories told in each chapter emphasise the local and the personal: narratives of specific people actually counting, for specific reasons. Some illustrate novelties and turning points, but most are about the way things usually were, the kind of events so common they are seldom written about or remembered. Within each chapter, the illustrations are often arranged by their date, but they are still branches on a tree, not stops on a highway, with 'later' often meaning different but seldom meaning better (or worse).

The story of counting has the tree's property that a closer look always brings more structure into view. That property comes to a head in the Americas. They were the last major landmass to be populated, and their languages and cultures are famously diverse, with dozens of different human groups across millions of square kilometres. Ways of counting in the Americas span the whole range from beads to tallies to words to symbols, with no clear, continent-wide preference. So by way of an epilogue, this final chapter presents something more like a tree in its own right, a microcosm of the world's counting in a 15,000-kilometre journey from the Alaskan Arctic to the Amazon basin.

But first, the roots.

PART 1

Roots

Number sense before counting

Can animals count? Do humans inherit from animal ancestors a 'sense of number', or even something more? The answer is a complex yes-and-no. Even the most gifted animals cannot learn to wield number words or number symbols beyond the first few, to perform calculations or to work an abacus. Yet many species do display a pair of abilities related to counting.

On the one hand, there is an ability to estimate *which is more* of two groups of items. The items might be chunks of food, predators, or members of the animal's own species; they might even be sounds or taps on the head rather than visible objects. The ability to make these kinds of judgements shows some consistent properties – and limitations – across the different species in which it has been found.

Well-designed experiments with humans, suppressing the more sophisticated ways of counting that nearly all have access to, can show that the ability to estimate in this sense is also present in *Homo sapiens*. You, too, can judge which is the larger flock of birds or the more numerous plate of cookies without actually *counting* them: and you can still do so when confounding factors like the shape or density of the flock are controlled for. This, surely, is one of the innate abilities that humans build on when they count in Leibniz's sense of paying repeated attention and keeping track.

On the other hand, most humans also share a sense that for very small numbers – up to about four – recognition is both immediate and exact. If you see three sheep in a field, you *just know* there are three: it doesn't feel like an estimation process, but it doesn't feel like a counting process either. It is more like pattern recognition, working at a glance: but it functions even if the objects are not presented in any special pattern or arrangement.

So it has often been suggested that there is another innate ability that sits alongside estimation and deals specifically with the smallest numbers. Sometimes called *subitising*, because it happens *subito*, suddenly, it has long been dogged by controversy, with some experts unconvinced that the evidence proves it exists at all. Experiments resist replication; results can be explained in more than one way. Perhaps subitising is just what estimating looks like when the quantities are small. If it is real, though, it is another ability that underlies human practices of counting the world over, and that can do something to explain why those practices have the characteristics they do.

These two capacities might be called proto-counting. Humans inherit them from the distant evolutionary past, and they lie at the root of what human beings do when they count. Although they are about animals, they are an important part of the human story of counting.

Estimating and the approximate number system

The Cayo Santiago, Puerto Rico, 1999: a tree-filled island amid the Caribbean waters. A rhesus macaque, foraging for food, spots something unusual. Two humans have approached. Each displays a coloured, opaque bucket; tips it sideways to show it is empty; places it on the ground. The first human places slices of apple in the bucket while the monkey watches; the second does the same. Both humans then turn and walk away. And wait.

After a few moments, the macaque goes to investigate. The humans watch, observe, record. The macaque can't see the contents of the buckets from any distance. But still it approaches, for preference, the bucket into which it has seen more pieces of fruit being placed.

Experiments like this one have been repeated with many species, and with similar results. It is not just monkeys that show a sense of number. Mealworm beetles can distinguish between different numbers of potential mates; cuttlefish can tell one prey item from two, two from three, and so on up to at least five. Certain spider species show a preference for settling with just one of their kind, rather than with none or with two or three. From frogs that count the pulses in their croaks to guppies that choose the larger shoal to swim with, and from parrots that select the greater number of food items to African elephants that can learn to choose between stimuli consisting of up to ten elements, something like counting seems to be everywhere in the animal world, present in nearly every species that has been tested for it. Cuttlefish, salamanders, barn owls, domestic chickens, New Zealand robins, pigeons, rats, bears, lions, hyenas, dogs, wolves, a dozen different primates . . . Forest, ocean and savannah seem to be teeming with numbers.

Something like counting can exist, then, without language, and without much – for some species, without any – training. Without a large brain; without a vertebrate nervous system.

Something like counting, but not really counting itself. The right word might be estimating; the technical term often used to describe animals' judgement of numbers is the *approximate number system*. What it does not provide is precision. It shows – and this is the same in every species tested – a characteristic pattern of errors, with discrimination becoming less accurate as the quantities get bigger. Rhesus monkeys can tell one from two, two from three, three from four, four from five . . . but start to fail from five upwards. Rats that learned to press a lever a given number of times, from four up to twenty-four, became markedly less and less precise in their responses as the number increased: by the top end of the range they would merely produce a spread of numbers around the target. It is a common observation that when testing the accuracy of animals' number sense, the size of the numbers matters.

In the same way, for the purpose of telling one number from another, the distance between them also matters: responses are always faster and more accurate if the difference is larger. Two and four are easier to tell apart than two and three.

Putting these two effects together, the best description of the approximate number system is that it is governed by a ratio. Most species seem to have a ratio above which they can reliably tell one number from another, while below it they become rapidly less accurate. For fish, the ratio is about two to one: so, for example, they can tell fifty objects from twenty-five, or two hundred from one hundred. Dogs and crows can do rather better and are accurate down to ratios of about three to two; some birds do better still with a limiting ratio of four to three. For rhesus monkeys, the ratio is perhaps six to five: they can tell, for instance, twelve items from ten or twenty-four from twenty. Estimates vary, and much depends on the type of task being carried out and, of course, the amount of training the animals have received. And different individuals are better or worse at it than others. An experiment with zebrafish found that of eight fish tested, some could only tell three from two, but others learned to tell four from three or even five from four.

One way to think of this is that if animals have anything like a mental 'number line', it does not have the numbers evenly spaced. Instead, the smaller numbers are widely separated, but the larger ones are more and more crowded together and hard to distinguish. No species on Earth can tell one hundred items from one hundred and one.

—·—

It is natural to wonder whether results like these are real. There have, after all, been some notorious hoaxes in the field of 'clever animals': in the world of animal numeracy it is best not to mention Clever Hans, the German wonder horse who amazed the world in the 1890s with his accurate responses to arithmetical questions. Last of a long line of calculating horses (there is an illustration from 1594 of a 'bay horse in a trance' doing arithmetic, possibly the same animal referred to as the 'dancing horse' in Shakespeare's *Love's Labour's Lost*), Hans naturally turned out to be responding to prompts from his trainer and questioner, who would move his head or his back when the right answer was in view. Hans and his like cast a long shadow over serious studies of number abilities in animals.

But recent experimental work is of quite a different kind, and it does seem that the animals really do recognise different numbers of objects as is claimed. It is easy enough to run experiments 'blind', with the experimenters unable to see the stimuli or barred from communicating with the animals. It is harder, but still in most cases possible, to make sure the animals are not responding to non-numerical cues, such as the total amount of foodstuff rather than the actual number of items (hence the business with opaque buckets for the macaques' fruit slices). Indeed, some animal species do fail when such controls are put in place: cats, for instance, in one study turned out to be relying on the visual cue of total surface area when choosing between two quantities of food, not on the number of separate food items. (The same was true for lizards, a group not noted for its number sense.) Salamanders seemed to be assessing the amount of movement they could see, not the number of fruit flies presented to them.

Despite such failures, for the majority of species tested the results stand: even when density, shape, area and arrangement are controlled or randomised, animals continue to be capable of selecting the larger number of items. Indeed, some will do so spontaneously, even when the task does not require it. There is evidence of spontaneous assessment of number in chicks, in dogs, and perhaps in a few other species as well. There are circumstances in which number is positively preferred over other parameters when animals make a choice. Chicks, for instance, if spatial and numerical cues conflict, will follow the numerical information. So will monkeys, preferring number to colour, surface area or shape as the basis for a choice.

Finally, the assessment of number by animals is not only a visual experience. Monkeys can be trained to compare a sequence of sounds to a set of visual items and correctly choose the visual array that matches the number of sounds. They are just as accurate at this as when they match visual with visual stimuli. So – approximate and fuzzy though it is – number seems to be not simply widespread and spontaneous in the animal world, but strikingly abstract too. It has been argued that number is one of the primary features of the world as many animals experience it, guiding decisions and forming part of the information an animal's brain encodes about a situation automatically, 'just in case'.

————

Why do animals have an approximate number sense? Traits evolve because they confer an advantage; because individuals that have them are more likely to survive, to thrive, and ultimately to reproduce. It is fairly easy to imagine situations in which natural selection would favour individuals who behave in line with judgements of number: those who join the larger rather than the smaller group of their own species, becoming safer from predators as a result. Those who flee from the predators only if they are numerous enough to pose a real threat. Those who choose the larger pile of nuts to forage in. Those who flock with the larger group of potential mates. Getting any of those judgements right will – for certain species – confer selective advantage in the long run. Many of these behaviours have been observed in the wild: in untrained animals, and in completely natural settings.

In the more aggressive world of carnivore societies, lions, hyenas and wolves are all thought to assess how many members of their own species they hear calling, and decide whether to respond or not depending on whether they have a numerical advantage over the rival group. A classic experiment demonstrated this by playing recordings of intruders to female lions and tracking their decisions whether to respond. Chimpanzees, too, have been observed to attack a neighbouring band only if the numbers are on their side. This ability to assess numerical odds correctly – defenders *versus* intruders – may be crucial for a group's success, and it is perfectly plausible that it is selected for in the long run.

In all of these situations, interestingly, it makes sense to be more accurate about small numbers than large ones. The difference between one and two apples matters more than that between twenty and twenty-one. Being outnumbered two to three matters more than by eleven to ten. The difference between a shoal of five and one of three is important; that between shoals of fifteen and sixteen is negligible. Natural selection is a plausible explanation not just for the presence of the approximate number system in many species, but for the fact that it is approximate rather than exact.

Estimation and humans

The University of Tübingen, Germany, 2008; a laboratory. Electric light on wood and plastic. A human being – let's call her Miriam – is being tested. She sits in a chair and looks at a computer display. It shows her a pattern of dots, then another, flashing each one up too fast to count. Do the patterns contain the same number of dots, or not?

Miriam completes dozens of repetitions of the task, and much of the time her answers are correct. She is more likely to be wrong when the number of dots is larger, or if the difference between the two sets of dots is small. Her ability to estimate numbers, in other words, turns out to be just like that of any other primate.

The approximate number system also exists in humans, although its operation is sometimes harder to see, overlain as it often is by a deeply learned impulse to count precisely using words, symbols, or some other device. If explicit counting is made impossible, though, estimation comes into its own. If you glance at a flock of twenty birds and a flock of thirty, you are likely to know that one is bigger than the other, even if there is no time to count them or even form an estimate of the actual number. In the lab, counting can be suppressed by keeping the stimuli brief, or by giving subjects tasks that interfere with verbal counting, such as asking them to read aloud at the same time.

Estimation. You know which flock has more birds
without counting them.

Under those conditions, the same features show up as in animals. Humans have a sense of *how many*, and it is an approximate sense: it deals in estimates, and those estimates become less accurate as the numbers become larger or the ratios between them become smaller. Everyone can tell two objects from three, or twenty from thirty. No one can tell a hundred objects from a hundred and one without explicitly counting them. Compared with other species, though, humans perform rather well. While even the best monkeys' limit of discrimination is a ratio of around six to five, adult humans have been observed managing ratios as small as eight to seven or even ten to nine.

As in animals, the human ability to estimate number is robust when experimenters control for confounding factors like the size or the density of the stimulus. It exists across cultures, and it works not only on the numbers of objects seen but on the numbers of sounds heard or physical taps felt.

An advantage for experimenters working with humans is that they can be asked more complicated questions and set more complex tasks than are feasible with animals. Thus in some ways the human approximate number system is better described than that in animals, with a dense tangle of different – occasionally contradictory – experiments and results. Estimation has an upper limit, for instance, beyond which a set of visible objects is perceived as having not a number but a texture. It shows, intriguingly, an effect called adaptation, shared with other types of sensory stimulus. If you put your hand in warm water, other objects will feel colder than they really are for the next few minutes. In the same way, if you look at a collection of a hundred dots, even for less than a second, then other, smaller collections of dots will look less numerous than they really are for the next few minutes. You will underestimate their number by up to a factor of two. The same effect occurs in reverse: if you put your hand in cold water, things will feel warmer than they are for the next few minutes; and if you look at a group of ten dots, other, larger collections will appear more numerous than they really are, until the effect wears off. The midpoint, the set of dots that is apparently 'normal' enough that it doesn't affect the perception of subsequent sets one way or the other, is around fifty. Rather wonderfully, this effect, too, is not limited to the visual mode of perception. In one experiment, subjects were asked to tap their fingers either quickly or slowly, and then to judge the number in a sequence of flashes or an array of dots. Slow tapping caused them to overestimate, quick tapping to underestimate.

The sense of number is already present a few hours after birth, albeit in only quite a fuzzy form, distinguishing, for instance, three items from one. But by four days old, babies can distinguish three-syllable words from two-syllable ones. The ratio of discrimination continues to narrow throughout childhood. And like animals, humans seem to form judgements about number spontaneously, even if they are not needed, and even if they actually interfere with the task at hand. People can't help processing the world in terms of *how many*, any more than they can help processing it in terms of colour and shape. Despite there being no obvious sense organ devoted to number, it is just as much one of the senses as vision, hearing and smell. Different people have it to a different degree, and those differences are at least in part genetically determined; one study with twins found around 30 per cent of variance in the approximate number sense was inherited.

———

Such a distinct, well-defined ability should surely have a dedicated part of the brain to perform it, although the fact that it is found in such a wide range of animal species raises questions about whether they can possibly all use comparable brain regions for the purpose. The classic technique of functional magnetic resonance imaging – which can show in real time which areas of the brain are being activated – has progressively narrowed the region in which approximate number processing takes place: to the neocortex (the outer layer of the brain); to the parietal lobes (upper back areas of the neocortex); to the interparietal sulci (grooves running along the side); and finally to the horizontal segments of the interparietal sulci. Reliably, in people from different cultures, adults as well as children, this is the specific brain region that first activates when numerical information is being extracted from the world, even before it is converted into words or arithmetic is done with the numbers: tasks which use different distinct parts of the brain, and which are indeed culturally dependent, depending on whether you do arithmetic using – or imagining – spoken words, written symbols, or an abacus, for instance. Numbers presented in any format – including as spoken words or as written number symbols – activate this same part of the brain.

There are similar results for animals, with monkeys also using a region of the interparietal sulcus, and crows a specific part of their very differently organised brains. It was long predicted, based on simulations, that there would be individual brain cells or clusters of cells specialised for the detection of different numbers. In 2002, a team successfully located individual cells in the brains of macaques whose firing was associated with the number of elements in a visual display. Certain cells were indeed associated with certain numbers. Present the monkey with two objects, and one set of brain cells would fire. Present it with three, and a different set fired. Even completely untrained animals turned out to have such 'number neurons'. In 2015, similar cells were found in the brains of crows, increasing their activity in response to numerical stimuli, and responding preferentially to particular numbers.

The number neurons turn out to do just what would be expected if they are responsible for the approximate number sense. They respond only to numerical information, not to other features such as whether the number is presented visually or as sound or rhythm. And they work approximately. Each one is tuned to respond to a particular number – say four – but also responds more weakly to neighbouring numbers on either side – say three, five and six. The precision of the tuning decreases as the numbers get larger: smaller numbers have more precisely tuned neurons, larger ones more fuzzily tuned ones, creating the possibility for making mistakes, and accounting for the observed limits on animals' abilities to discriminate between a number and its neighbours.

It is an exciting set of results, and it confirms what had been suspected from neural network simulations: that you do not need a large brain or a large set of dedicated brain cells to detect numbers in your environment. Computer simulations involving as few as twenty-five dedicated cells have been shown to do it successfully, and with a few hundred cells some of the more complex features of extracting number from a visual display can be replicated. Simulations also hint that forming representations of numbers is something a network of a few hundred brain cells may start to do spontaneously, given the right combination of input and reinforcement: in other words, it is an easy capacity to acquire, a potential that even the simplest brains might have.

———

So it looks very much as though humans inherit from their animal ancestors an ability to detect, and encode approximately, numerical information from the environment, from a range of different types of stimulus. But it is not yet clear just how ancient the ability is.

The fact that humans and macaques use the same part of their brains for the approximate number sense suggests very strongly that the ability was present in the common ancestor of primates, about twenty-five million years ago, and was passed down to humans through the line of pre-human hominids. The data for other mammals are too incomplete to make any similar claim about their common ancestor. But the fact that the approximate number sense turns up with a relatively high degree of precision – comparable to that of monkeys – in some birds, is intriguing. The last common ancestor of birds and mammals – a reptile-like animal – lived about 320 million years ago. The brains of the two lineages have evolved separately ever since, though to some degree in parallel: the bird brain is of quite a different design from a mammal's, having evolved for low weight (they have nearly twice as many neurons as a primate brain of the same mass), despite providing many similar functions in response to similar pressures from the environment. It is conceivable that that common ancestor could already detect and distinguish quantities, but perhaps rather more plausible that the approximate number sense has evolved independently in the two lineages. This suggestion is strengthened by the limited data for reptiles, which usually seem to have more restricted numerical abilities; it indicates, perhaps, that birds acquired their numerical skills after they split from the dinosaurs around 150 million years ago.

And it is the same story with the fish – and still more so the invertebrates – that have shown numerical abilities. Although it has been suggested that numerical abilities go right back to a common ancestor half a billion years ago, it is more often postulated that the different groups have evolved their numerical estimation abilities independently: testament as much as anything to the strength and constant presence of the environmental pressures that make it useful to have an approximate sense of number.

At a glance: Subitising

A curious pendant to the approximate number sense is the way people can recognise small numbers at a glance. Some experiments seem to show this as a separate ability from estimation; collections of one, two, three and four items appear to be distinguished immediately, without the effects of size and ratio that characterise the approximate number sense. Babies just a few hours or days old, for instance, have been observed to tell apart collections of two and three objects but not collections of four and six. At a few weeks, they can tell apart sets of one, two, three or four items, all equally fast and equally accurately. But their performance drops sharply if the number of objects becomes larger. For adults, the limit of this ability – 'subitising' – seems consistently to be four, with the difference between three and four recognised just as reliably as that between one and four, and the ratio effects characteristic of the approximate number sense appearing only when the numbers are larger than this.

Like the approximate number sense, subitising is cross-modal: heard or even felt stimuli can be shown to be distinguished in a similar way. (Many readers will have experienced the difference between counting musical beats and 'feeling' them, which surely has something to do with the limits of subitising in the rhythmic domain.)

Subitising. You know there are three sheep without
thinking 'one . . . two . . . three'.

But, in another difference compared with the approximate
number system, subitising seems to stop working when your
attention is distracted. While judgements of approximate number
are formed 'in the background' of other tasks, subitising requires
you to focus on the objects in order to know how many of them
there are.

Subitising, in fact, seems to be very much about tracking and
paying attention to the individual objects. Many researchers have
suggested it works something like a set of slots or pigeonholes in
your working memory: you can fill them one at a time, but when
all the slots are full, you lose track and fall back on estimation.
For that reason, this particular ability is sometimes called the 'object
file system'.

Within its range it can even provide for some basic arithmetic.
In a classic quip, two tigers go into a cave and one comes out. Is
the cave safe? If you are keeping track of the tigers as individuals,

you know that there is still one in the cave, even if you have never learned such a thing as *two minus one equals one*.

———

It is fair to emphasise again that subitising is controversial, and little about it is really settled. One intriguing question is whether subitising actually replaces estimation for small numbers, or whether instead both systems work alongside one another for object collections numbering from one to four. The object file system can be turned off by distracting the attention, and in that case most studies show that estimates of numerosity are still successfully formed, and that the ratio and size effects characteristic of the approximate number system can be discerned again. Does that mean that approximation is the more fundamental system, overlain in some situations by an ability to track up to four objects precisely? Or, instead, does the approximate number system normally cease to function when the object file system is engaged? The answer is not yet clear.

Another question is whether subitising is possessed by any animal species. It would be surprising if such an ability existed in humans only, but the evidence for it in animals is patchy and in some cases problematic. Some would argue that for certain species the data are best explained by a subitising-like process. Bees, for instance, have been reported to distinguish successfully between sets of up to three objects, while dropping abruptly to the level of random guessing when the sets are any larger. That is the signature of an object file system, not an approximate number system. The same is true for salamanders, while the abilities of certain fish species look more like the two systems of humans: high accuracy for collections of up to three objects, succeeded by an increasingly fuzzy ability to approximate for larger numbers. The evidence for birds and mammals points in various directions. Subitising has been reported for robins, chicks, pigeons and parrots, for dogs, monkeys and chimpanzees, and some scientists certainly regard it as widespread in the vertebrate family tree, and therefore presumably of some evolutionary age. But retesting has not always succeeded in replicating these effects, and much uncertainty remains.

Recent reviews of the evidence emphasise that nearly all animal species tested show signs of an approximate number sense, whereas the evidence for a second system remains limited and inconsistent. Some would say that approximation, which is at its quickest and most accurate for small numbers, can account for all the observations yet published. Others would add that it is not straightforward to think of factors in the environment that would create selection pressure specifically for a subitising-style recognition of small numbers instead of – or as well as – an approximate system. One researcher speaks of an 'impasse in the literature' on these questions.

———

So, can animals count? Yes and no: but mostly no. Animals do not count in the sense in which (most) human beings do. Despite its capabilities, and despite its great age, the approximate number system is not a counting system. There are no exact numbers here: just approximations that become rapidly more fuzzy. This is a system in which one and two are distinct, but three and four less so; and in which one hundred and one hundred and one are for all purposes identical. Unlike counting, when every step from a number to the next is the same size, the approximate number sense is a system in which the relationship between two and three is very different from that between, say, fifteen and sixteen.

Similarly, subitisation – if it is real – seems to have more in common with pattern recognition than with counting: it is a form of perception in which a group of two objects and a group of three are sharply distinct, but which has nothing whatever to say about the difference between groups of four objects and five. If it is right to think of it as an object file system – a set of mental slots to be filled – then it really contains no explicit representations of *how many*, and the fact that you can get a number from it is a mere side effect.

These innate abilities are roots (seeds, even?) of real use for constructing counting practices, but they are not the same as counting. They constrain what counting can be like; they constrain which species can acquire it. But there remains a real gap between these abilities and counting. Humans are not born able to count, then; they have to learn to do it. Cultures have to invent ways of counting and transmit them from one generation to another, with the real possibility that ways of counting may change out of recognition over time, or die out or be replaced.

2

Counting before writing:
Africa and beyond

If inherited abilities are one root of human counting, another is those things in the environment that have sometimes been called the 'wild number lines'. Things like pebbles that you can use as counters, butchery marks that you can repurpose as tallies, or the set of fingers and toes that (nearly) all humans have. Each can be used to keep track of a sequence and thereby make counting possible. Each can serve the functions of a set of numbers, recording and communicating the outcome of a count. And so, of course, can words, once humans have started to use them.

Each of these four wild number lines is grounded in Africa, like the human species itself. For counters and for tally marks, African archaeology provides some of the earliest evidence, and therefore a tantalising glimpse of these roots of counting in the Stone Age. For counting on the fingers, the best early evidence that survives is from north of the Mediterranean. For words, there is no really direct evidence at all, although much can be deduced from counting words in living languages. But the story starts with beads.

Blombos: Counting with beads

Blombos Cave in South Africa, 75,000 years ago. A woman takes a shell, collected from the coast. Much the same size and colour as a large human tooth, it was once the home of a small sea snail. Now it will be a bead.

She pierces the lip of the shell with a bone point, lays the shell aside and takes up another. When she has enough, she threads them on a strip of hide or plant fibre; ties it. A necklace.

The human and chimpanzee lineages split around seven million years ago. Over a dozen early hominid species are now recognised, and the shape of the family tree changes as new fossil material is uncovered. About eight species, at different times, made up the genus *Homo*, which evolved in Africa over two or three million years. These species committed to living on the ground and grew a proportionately enormous brain compared with their ancestors. Brains do not fossilise, but there is some evidence from the shape of fossil skulls that the parts that grew the most included the interparietal sulcus, used for the approximate number system, among many other functions. More visible in the archaeological record are changes in behaviour: butchery marks appeared on bone as early as 3.2 million years ago, and stones modified for cutting first appeared in the Kenyan and Ethiopian Rift Valley 2.6 million years ago. Fire was used, perhaps, as early as 1.5 million years ago.

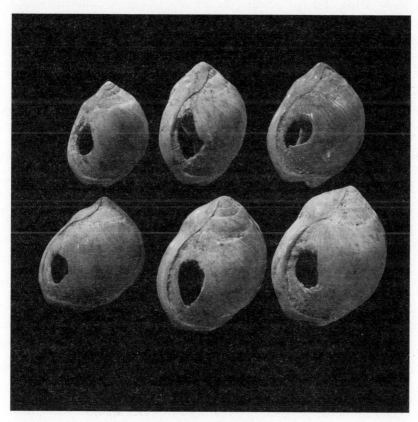

Beads from Blombos.

The period called the Middle Stone Age began perhaps a quarter of a million years ago in Africa. It was distinguished by a new, more sophisticated suite of stone tools: flakes and blades, cores and points. This was a more variable set of tools than its predecessors, but still it lasted for many tens of thousands of years without a great deal of change. The archaeological record is coarse-grained, though, and the scales of time and distance involved are enormous. There were anatomically modern humans – *Homo sapiens* – in Africa by 200,000 years before the present, perhaps by 300,000 years; by 100,000 years ago, the species could be found from South Africa to the Levant. Lifestyles were based on hunting and gathering. Shelter was in caves, many of which are still visible.

Blombos Cave is located in the southern Cape, South Africa; it is about 300 kilometres east of Cape Town. Today, the Indian Ocean is just 100 metres away. The cave site was occupied in three or four phases, starting upwards of 100,000 years ago and ending perhaps 70,000 years ago; the occupations may have been relatively short in relation to the millennia that separate them. Beach sand seals the last Stone Age layer.

Those who last lived here made a range of artefacts including stone points and bone tools. They made much use of marine resources such as seals, dolphins, fish and shellfish, although the sea level was lower in their time and the coast certainly further away: at times, dozens of kilometres. The climate was relatively warm and moist during the final period of occupation seventy-odd millennia ago; the landscape a patchwork of open areas, shrubs, trees, waterways and woodland, constantly changing under pressure from plants, animals and slowly shifting rainfall patterns. The people also moved around, to forage, to hunt and to sleep.

The cave at Blombos is not a large one, and no more than perhaps thirty people would have stayed there at any one time. But its excavation since 1991 has revealed a much-studied set of artefacts in addition to what is usual for its time and place. There are thousands of pieces of ochre, including some bearing unmistakable signs of engraving: this at a very early date for abstract representations anywhere. And there are beads.

The Blombos beads are, so far, the oldest Stone Age beads to be dated really securely; indeed, they are the oldest securely dated

personal ornaments of any kind. They date from the last occupation level at Blombos Cave, that is roughly seventy-two to
seventy-five millennia before the present. The beads are made of
tick shells: the species is called *Nassarius kraussianus* and lives only
in estuaries. This was an exotic material and an expensive one in
terms of time and effort; the shells would have had to be trans-
ported from rivers twenty kilometres or more from the cave.

Sixty-eight tick shells have been found at Blombos. Nearly nine
in ten have a hole near their lip, made by inserting a bone tool
through the main opening and then pressing. This was the action
that transformed them into beads. Once pierced, the beads were
threaded onto cord or gut; traces of wear on the edges of the holes,
and on the outer sides of the shells, show where they rubbed against
both the cord and each other. There could be as few as two beads
on a necklace, or as many as two dozen; up to about twelve was
average. They were strung in symmetrical pairs, one facing left,
one facing right. The marks are deep: the strings of shells were
worn for months or years.

Shell beads have also been found at a range of other archaeological sites, with some possible dates – less certain than at Blombos
– as old as 115 millennia. Most of the sites are in south or east
Africa, some in Morocco and one or two in Israel. A common
pattern is that the shells were from species exotic to the locations
where they were found; they were brought in from some distance
away. Particular species and sometimes particular sizes were selected.
Sometimes natural holes from wave or beach action were used for
stringing; other beads, like those at Blombos, were deliberately
pierced. The tradition of shell beads lasted many thousands of years
and ranged over five or six thousand kilometres; at one time it
may have been common, beadwork one of the everyday items of
human life.

In archaeological terms, beads are one of the first visible examples of symbolic behaviour, something like evidence for culture

rather than mere survival. Are beads also evidence of the first steps towards human counting?

Here the mystery deepens. The beads are from millennia before writing; if they had a meaning, their archaeological context cannot show what it was. Beads were surely first made as personal decorations, with a purely aesthetic significance. But they are probably also some of the first surviving objects that were meant to communicate information – information about their wearers and their communities – making them some of the first evidence for humans' manufacture and use of symbols.

Beads also have a suggestive set of properties that have led more than one scholar to describe them as a wild number line – capable, in the absence of number words or gestures or even number concepts, of existing and doing some of the things that counting sequences do. If you see a modern pre-schooler playing with beads on a string, you assume she's learning important things about sequence, order, and eventually number. About adding-one-more, or creating a new sequence by joining two shorter ones together. About subtracting from a sequence to get a shorter one, or even dividing a sequence into equal portions.

If Stone Age Africans had similar devices in their hands, surely they learned similar things. With beads, you can play, manipulate and contemplate processes like sequencing, adding and subtracting as physical actions – perhaps for years, perhaps for generations – before eventually learning to associate a numerical meaning with them. Not only that, but compared with fingers or the small collections of objects that can be subitised, strings of beads can embody quite large numbers: up to twenty-four, in the case of the artefacts recovered at Blombos.

You don't necessarily even have to have a set of counting words in order to use the beads for counting: in order to correlate them one-to-one with other objects in the world, or with actions or events. There are, to this day, many situations in many cultures in which a ritual action is done once for every bead on a string: prayers or prostrations, for instance. No counting words need be uttered and no number concepts need be held in the mind; you simply hold one bead and perform the action, then you hold the next bead and perform the action again. When the beads run out you stop performing the action. You don't have to use or even know any number words in order to use a set of beads like this; indeed, it is not clear that your culture has to feature any other way of counting in order for a string of beads to work for this purpose.

In other words, it is possible to imagine a world in which the beads on strings are the main, the best, and even the only available way of counting. The beads at Blombos and other locations may have existed and been manipulated for many generations without anyone using them to count. But they may, eventually, have been used to keep track of things or objects – to count – even if the culture of which they were part contained no other way of counting. It is not certain that such a world ever existed, but the possibility is a most intriguing one.

Lake Rutanzige to Laussel:
Counting with tallies

Perhaps 20,000, or even 25,000 years before the present. A woman poses, holding up the horn of a bison. The mother of children, she is pregnant again, expecting her new child in a month or two, and with her free hand she gestures to her abdomen. She turns her head to her right, to look along the length of the engraved horn; as she does, her hair falls across her shoulder.

The horn itself is engraved with thirteen short parallel marks. A decoration? Perhaps. A count? Very probably. But what was she counting?

Something the Stone Age environment contained in large numbers was the bones of animals, many of them bearing scratches and cuts from butchery done with stone tools. During the Middle Stone Age, from as early as 90,000 years ago, some human groups also began to work bone, making bones into pointed tools and other artefacts.

The combination of accidental scratches and cuts on bones, together with the increasingly skilled working of bone, led some humans to mark bones with deliberate scratches; and, eventually, to use those marks as a way to record or communicate information. This, again, is a world of tantalising clues and ambiguous evidence;

once the original context is gone, it is seldom possible to be confident what information an artefact was meant to communicate. But, on the one hand, a series of neat, symmetric, cross-hatched incisions on a bone tool looks irresistibly like decoration, and raises questions as to what it might have conveyed about a person's identity, or a group's. And, on the other hand, a series of roughly parallel scratches on a bone (or other) surface looks irresistibly like a tally, the outcome of a counting process: one scratch for each object or each event in some set or sequence.

The site at Blombos provides, as well as shell beads, some of the most intriguing of the very early evidence for deliberate marking of bone. Working bone was, indeed, a regular activity here: twenty-eight bone tools – awls, spear points and a retoucher – have been recovered from the early archaeological deposits. The age of these bone artefacts is over 70,000 years, and at least two of them bear possible deliberate engravings. One has eleven incisions parallel to its long edge, while a more superficial oblique line intersects six of them. These are certainly not random marks, and microscopic analysis shows that they were produced by the same stone point in a single session. Their design was intended from the outset and made deliberately.

As with the bead strings from Blombos, these earliest marked surfaces most likely bore no numerical meaning, if indeed they carried any meaning at all. Scratches could have been made on bones for years or generations as an object of play and experiment before they meant anything. But making a series of marks on a surface is, at the same time, another 'wild number line': another artefact in the human environment capable of being eventually put to numerical use. And like the beads on a string, it could potentionally function as a way of counting even in a culture that contained no other way of doing so.

Archaeological evidence for patterns of deliberate scratches on bone and stone – and on shell – is present sporadically throughout the African Middle Stone Age, and towards the end of that period it becomes possible to be confident that counting was involved. Archaeologists emphasise the fact that a proportion of the marks and scratches that have been interpreted in these ways are the results of natural phenomena such as gnawing or of damage during excavation, or are interpretable as non-deliberate butchery marks rather than patterns made deliberately. They note that some patterns of scratches may have been made in order to make the surfaces of tools easier to grip. They acknowledge that meaningless doodling by Stone Age people is also a possibility, as is decoration without any specific information content.

But the consensus does seem to be that the step from decoration to symbolism was taken by Stone Age people, and that some of the marks on very ancient artefacts recorded information for the people who made them. That information could have served various different functions for the hunter-gatherer societies in which the artefacts were made, and some of it may well have been numerical information. In particular, where there is a sequence of similar marks in a repeated pattern, regularly spaced and running along a well-defined path, it is not unreasonable to see this as evidence of counting, and where there is subdivision of the marks into (irregular) groups, that possibility becomes all the stronger. Some scratches really are tallies.

For instance: between 20,000 and 25,000 years ago a group of humans lived by what is now Lake Rutanzige in the Democratic Republic of the Congo. Theirs was a more settled community, perhaps, than those of their predecessors, and they relied on spear and harpoon to hunt fish in the lake. They used quartz to tip their tools; one that survives is the fibula of a mammal about 10cm long, with a quartz fragment fixed in a cavity in one end. Its exact purpose as a tool is uncertain: engraving or tattooing are possibilities. It is probably the most-discussed artefact in prehistoric mathematics, however, because as well as its functional characteristics it bears sixteen groups of engraved lines, arranged in three columns. For what it is worth, the groups contain 11, 13, 17 and 19 marks in the first column, 3, 6, 4, 8, 10, 5, 5 and 7 in the second and 11, 21, 19 and 9 in the final column. These are well-defined, clearly grouped marks that can scarcely be interpreted as accidental or decorative. Interpretations of the bone have focused, understandably, on the assumption that these are tallies, made as part of a counting process and in their finished state communicating the end-points of the counts to others. Potentially, communicating them to the present day.

The Ishango bone.

But what were the makers of this artefact counting? It seems impossible to say. What possessions might have needed counting? What resources managed? What events tracked across time? One hypothesis is that you might need to count to keep track of the traps you have set for animals. Another is that you need to count food items in order to ensure survival over the winter, particularly if the region's climate is becoming colder or drier. Trade, gambling or ritual are other possibilities, and the distribution of surplus at feasts is another. But all of these ideas make assumptions about the society in question that no one can test.

Much ingenuity has been exercised in order to discern some regularity in the tallies, in the hope that this will reveal something about their meaning. There are 168 marks in total – sixty in each of the outer columns and forty-eight in the middle one – and it is tempting to imagine an attempt to divide up some commodity that came in twelves. Other readings focus on the presence of prime numbers only in the first column, pairs of numbers bracketing ten and twenty in the third, and relationships of multiplication by two in the middle column. It is not impossible that these counts arose from playing a game of some kind; a calendar, on the other hand, seems hard to fit to these numbers, though it has been tried.

The bone from Lake Rutanzige (it is now normally known as the Ishango bone) illustrates the sophistication that African tallies could have by the later Stone Age; it also illustrates the mysteries that will always surround this kind of evidence. It seems at least reasonably clear that the bone comes from a situation in which the individual marks meant something, that the point of the engraving was not the pattern as a whole but its details. But certainty that these marks were intended to represent numbers is elusive; confidence about what the numbers counted is quite unavailable.

Homo erectus spread out of Africa into Eurasia more than a million years ago, and by half a million years before the present, tool-making hominids could be found from the tip of Africa to Spain, England and northern India. From perhaps 100,000 years ago waves of modern humans (*Homo sapiens*) began to leave Africa: some successfully, some not. Some travelled along the coast into Asia, others through the Levant into Europe. Eventually their descendants swamped the *Homo erectus* descendants they found – Denisovans in Asia, Neanderthals in Europe, whose capabilities and cultural sophistication are a matter of debate – and spread further: to China, Australia, Japan, Indonesia, eastern Russia, North America, South America, as well as migrating back into North Africa. Humans had reached Australia by possibly 65,000 years ago, and dominated Europe from perhaps 40,000 years ago.

The trickle of tallies and possible tallies becomes a much stronger flow during the later period, especially in the prehistoric archaeology of Europe. Artefacts of this kind have been found at over forty different sites, enough to give an impression of a widespread and perhaps continuous tradition over a long span of time. Nearly all of the marked objects are small; many are hand-sized or smaller portable objects, and their survival and discovery are a matter of chance. It seems highly likely that they are the remnant of a much larger number of objects that once existed, and which very possibly included engravings on perishable materials such as wood, which do not survive in the archaeological record.

For the objects that do survive, microscopic study and painstaking replication by modern experimenters have yielded a wealth of information about how the marks were made. Lines could be made with a single stroke or with multiple strokes, 'notches' made by moving the cutting edge to and fro, 'microcups' by rotating a point. Tools could wear down, break or be replaced; they could be shifted in the hand or resharpened. These marks record in an unusually direct way the individual actions of those who made them: the individual strokes of a blade on a bone or another surface. They give the irresistible sense of bringing the thoughts of prehistoric people close to the present.

Yet those thoughts remain tantalisingly elusive, because the reason why the marks were made remains uncertain. The history of the modern answers to that question would, indeed, be a study in its own right. The first archaeologists to discover such objects during the nineteenth century frequently read them as counts of animals taken in the hunt. That view gave way during the following century to a plethora of different interpretations, including that they were records of the numbers of participants in rituals or for use in games. Popular in the 1980s was the reading of sets of scratches or notches as lunar calendars, on the analogy of tally calendars from native North America. One after another, these readings have fallen victim to a lack of direct evidence, and to the fact that once it is removed from its original situation, a symbol may mean almost anything.

Occasionally, though, there is more context to help. Palaeolithic Europe was a place and time where cave art – representations of people and animals drawn on walls – had begun to appear. In a rock shelter at Laussel in the Dordogne, people engraved a series of human figures into the limestone. One has now fallen to the ground (or been 'excavated' by the heroic methods of the nineteenth century) and been subsequently removed. Not all of the figures can be easily read, but there are certainly two women among the group as well as an adolescent, and – the one now removed – a pregnant woman holding a bison horn in her right hand.

This figure bears what is very probably a tally: the horn has a series of thirteen marks on it. The figure is about 40 centimetres high, and traces of red pigment – ochre – indicate that it was once coloured in whole or in part, perhaps mimicking the use of ochre on actual human bodies in this culture. The period famously produced a large number of depictions of women, notably in the form of small portable statuettes. Compared with later ice age art they are not particularly schematic: indeed, some archaeologists have felt that their details make them portraits of individuals rather than representations of ideal types, still less of goddesses (the old designation 'Venus' still lingers around some of them). At the least, they realistically represent different physiological situations such as youth, age or pregnancy. To this end, they tend to neglect details such as hands, feet and hair.

The woman of Laussel.

The woman of Laussel is a superb specimen of this art. Unusually, her hair is clearly depicted; as is one of her hands, complete with its fingers, which gestures towards her abdomen as though to draw attention to her pregnancy. Collarbone, hips and navel are all clearly drawn or suggested, and her pose is realistic, with one shoulder slightly higher than the other, because of the movement of the arm to hold the horn. She turns her head to look along the length of the horn. On the other hand, her face is not even sketched, and her feet are hardly suggested; and the proportions of the body have been somewhat distorted in order to make the pose work. She looks, for all that, more like a real person than the figment of an artist's imagination.

The marks on the horn she holds seem to form a single series. Unlike those on the Ishango bone, there is no internal structure provided by grouping of the marks: that is, the gaps between them are all about the same size. The marks themselves are not quite identical; the second from the end is more like a Y than a single stroke (there is a mark of similar shape on the woman's left hip), and differences in length and width might make the subsequent eleven into two groups of four and one of three. But this may be to over-interpret the image.

Who was she? What does the image mean? And what was she counting? As with prehistoric art generally, there have been many attempts to interpret the woman of Laussel, ranging from a record of animals hunted to a magical icon; perhaps even a request for a certain number of animals. The horn has been seen as a drinking horn, a musical horn or a horn of plenty. A reading as a 'lunar calendar' has been tried, too: thirteen – the number of marks – is after all the number of lunar months in a year, and the horn does look something like a crescent moon. Recent interpretations have circled around the idea of a fertility calendar or an obstetric calendar, linking the woman's obvious pregnancy to the possibility that the horn reports a count of days or months. As the museum of Aquitaine, where the figure is held today, puts it, 'unfortunately, the deeper significance of this art will probably remain unknown to us.' Prehistoric tally marks, indeed, will always tantalise.

Cosquer: Counting by hand

The north coast of the Mediterranean, 27,000 years ago. In the long slope between the ice and the sea, men and women enter a cave. They carry charcoal, ochre and pine-stick torches. A tunnel leads to a huge chamber, too big for the torches to light completely. Stalagmites, stalactites, pillars and concretions on the walls add to the weird, other-worldly feel of the place. The people climb and balance. One takes a mouthful of pigment, presses a hand high on the wall and sprays the pigment out, creating a stencil.

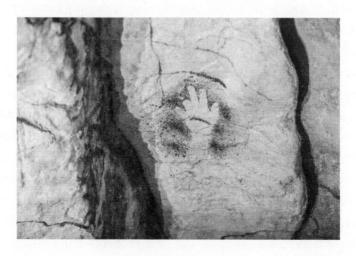

Handprints from Cosquer.

So the Stone Age environment included beads on strings and marks on surfaces, each of which could be made, used, experimented and played with both before and beyond the point when they were used to keep track of *how many* of something: used to count. It also had available the human hand, which provided a more limited but more immediate option for keeping track of processes or things.

Compared with speech, gesture was late to be recorded in writing, and late to be studied seriously by people interested in the ways and origins of language. But it is now regularly speculated that gestures actually preceded spoken language in the ancient human past. Certainly, gestures seem to be basic and obvious – both to oneself and others – and are made spontaneously by human children, even before they learn to control their vocal apparatus. (It is the same story with other primates, which generally show much more spontaneity and flexibility with their gestures than their vocalisations, and have much more success learning new ones.) Perhaps early hominids pointed and pantomimed through gestures – to direct attention, to gain desired behaviour from others and to facilitate cooperation – even before they acquired speech.

For the purpose of counting, or as a wild number line in existence before counting, the fingers form a sequence with a stable order: and they are always, literally, to hand. As with beads on strings or marks on surfaces, people could flex their fingers in sequence for many generations before beginning to use them to keep track of objects or events. Thus they are another possible bridge from playing and experimenting to counting.

Another advantage is that the digits of one hand take you just beyond the subitising limit of four. If the first four finger gestures can be distinguished and recognised using the subitising capacity, a single hand provides just one more item in the sequence: a whole-hand gesture easy to make and recognise. Perhaps this could have been a first step from the small numbers – recognisable at a glance – to what comes beyond them.

In a similar way, extending or flexing the fingers of one hand might act as a bridge from counting for oneself to communicating with others. A finished gesture does not persist the way a mark or an artefact does, but it can be shown to someone else, and perhaps become an answer to the question 'how many': a 'number word' in a certain sense.

To check how plausible these speculations are, it is not possible – unfortunately – to observe people today using gestures in isolation from other ways of counting: because there seems to be no culture that uses gesture as its only way to count. It is possible to observe young children, who learn early to extend one finger after another, and – in a modern, number-saturated environment – learn early to associate those gestures with other ways of counting. There is convincing evidence that children learn number gestures before they learn number words, and indeed that in the early stages of learning they may be more accurate when using gestures for small numbers than when using words: even that their acquisition of gestures precedes and facilitates the acquisition of words.

So widespread is finger counting, indeed, that its traces persist in adults' handling of number and calculation. Errors of size five – a whole hand – are more common in adults' arithmetic than errors of other sizes, even when words or written symbols are being used rather than fingers: which may suggest that people continue to associate numbers at least in part with hand gestures. Finger gnosia – the ability to tell your fingers apart when they are touched but you can't see them – is a good predictor of arithmetic ability: better in fact than general intelligence. And adults on average count and calculate more slowly if they are required to do something else with their fingers at the same time.

As well as this modern evidence, there are also – remarkably –

direct traces of the use of the fingers in the Palaeolithic, from certain human groups which found a way to make their gestures endure.

⸺

During the last ice age (roughly 30,000 to 10,000 years ago), a strip of land in what is now the southern fringe of Europe was occupied by humans. Although much of the continent was technically a polar desert, the inhabited area, beside the Mediterranean Sea, was not impossibly cold, with temperatures up to 12 or even 15° Celsius in the summer and relatively mild, survivable winters. Sheltered valleys had woodland with both conifers and broadleaved trees, and there was a range of land animals as well as maritime fauna: enough to support a human population. Comparable kinds of tools existed from northern Spain to the Russian plains – knives, arrows and spears – as did comparable artistic traditions. Little is certain about these people's lifestyle: it was heavily based on hunting, and they may have followed animal herds from place to place during the year.

Lifestyles were certainly shaped by the climate, and natural shelters were often used. One was a cave near the southern coast of what is now France, a few kilometres southeast of Marseilles. It opened at the foot of a limestone cliff with an entrance wide enough for two or three people to walk abreast. The entrance tunnel climbed gradually over a distance of 175 metres until it opened into a large chamber, 60 metres across and in places three times the height of a person. At the far end, there was a dome 30 metres high and a shaft nearly as deep. The slope of the land meant the chamber was far below the surface, and there was only one entrance. It was an isolated, mysterious place, and except for torchlight an utterly dark one.

No one lived there. Fires were lit to provide light, not to

cook. Torches were carried in; charcoal from them fell scattered on the floor. And the walls, by the flickering light, were decorated with paintings and engravings. The earliest were made more than 27,000 years ago.

Finger-marks were made in the soft calcite that covered the walls. People went to every part of the cave, even – perhaps especially – to the most inaccessible parts of the roof, covering dozens of square metres with simple lines made by the fingers. Curves, zigzags, scrolls intersecting one another. They formed a background showing – apart from anything else – that humans had been here, had made the place their own.

People also drew and scratched on the hard parts of the walls: pictures of a bison, perhaps of some horses. They drew quickly, in minutes, with simple schematic lines, designs up to a metre or more long. They added signs to some: long barbed lines superimposed on the animals, angles or chevrons, zigzag bands. Were they icons of the hunt? of weapons? or representations of animal footprints and faces?

Added to the walls around the same time – and involving much more effort, time and preparation than the simple schematic animals and signs – were dozens of stencilled hands. The technique was to dilute pigment – red clay or charcoal – with water, and blow it from the mouth, possibly using a tube. A hand pressed against the wall could thus make a striking stencilled image: a red or black halo with a clear, sharp image of a hand centred in it in the colour of the rock itself. If a single puff of pigment didn't work, it could be repeated, or the image retouched by dabbing on extra pigment by hand. The result was a dense ensemble of hands stencilled in various parts of the cave.

Stencilled hands are one of the iconic images from the art of the ice age, and they are known in three dozen caves across what are now Spain, France and Italy. They are familiar beyond that cultural area too, for instance in the art of Aboriginal Australia.

The French cave, now known as La Grotte Cosquer, has more than sixty hand stencils, one of the largest collections known.

The hands are distributed unevenly around the cave walls, exclusively on the eastern and northeastern sides of the cave: the right side as you go in. They are concentrated near the shaft and the vault at the eastern end of the main chamber: that is, in the more dangerous section of the cave. Many are in positions that would have required some climbing. There is a hint, perhaps, that a set of mainly red hands leads around the main chamber towards the shaft, and a large set of black ones stands beside the shaft itself, including on a stalagmite above it.

Most appear to be left hands (assuming the hand was held with the palm against the wall to stencil it). If the people were, like people today, mostly right-handed, perhaps they were using the more agile right hand to hold a torch or a supply of pigment, or to cling to a support when the position was a difficult one.

Whose hands were they? The stencilling process is a mechanical one; it records the characteristics of the maker's hand automatically, unaffected by questions of style or artistic convention. The hands bring close – in a sense, deceptively close – the individuals who made them: they are the signs of individual bodies, even of individual identities tens of thousands of years ago. They feel almost like signatures. The makers were adults: there are no child hand stencils at Cosquer. Some were over two metres tall, unless they stood on each other's shoulders or built scaffolding in the cave to reach the higher spots on the walls.

The hands stencilled were generally large and robust, though some of those by the shaft show a slimmer wrist; finger ratios and lengths are often – not always – consistent with those of modern women. Certainty is impossible, but both men and women seem to have been involved. There are what appear to be pairs of stencils from the same hand: one with an unusually long forefinger; another with a thicker ring finger. But just three of the stencils have been directly dated by archaeologists so far, and it is impossible to say how many people's hands are represented at Cosquer, or what period of time was involved. Is the stencil scheme the work of a single day? Of a lifetime? Or a record of visits to the cave by many people over hundreds of years, as mysterious to each other as they are to visitors today?

––––

And the evidence for counting? Some of the stencilled hands are incomplete. That is, certain of their fingers are shortened. Of the fully legible hand stencils at Cosquer, rather more than half, in fact, show less than a complete hand. These incomplete hands are, for some reason, much more common among the red stencils than the black. Incomplete hand stencils are known from a few other caves in France and Spain, but most of those known to exist are at Cosquer.

How do you produce a stencil of part of a hand? Suggested explanations have included that the hands in question were really incomplete: that these hunter-gatherers suffered from frostbite, from disease, or practised ritual mutilation. But their sheer number – and the impracticality of living a Palaeolithic life with a mutilated hand – count against these possibilities, and the balance of opinion seems to be that the simpler explanation is right: that the incomplete stencils were made by folding down certain fingers. Modern experiments have confirmed that it is possible to make stencils like those at Cosquer by folding down one or more fingers and pressing your hand against a rock surface. The bent fingers prevent the hand from quite touching the rock, resulting in a characteristically slightly fuzzy outline at those points: a characteristic that can be seen in the stencils at Cosquer.

Given four fingers and a thumb, there are in all thirty-two possibilities for folding down some, none or all of them. At Cosquer

just five of those possibilities were used, namely those patterns in
which the extended digits are thumb alone; thumb plus index
finger; thumb, index and middle; thumb, index, middle and ring;
and all five fingers. It is impossible that this was a matter of chance.
Moreover, one of the patterns – folding down just the little finger
– is relatively hard to make. This set of five hand shapes was deliberately chosen by the stencil makers at Cosquer.

Why? As with other prehistoric signs and proto-signs, the more
general question of the meaning of Palaeolithic hand stencils has
been a vexed one for over a century; one scholar aptly says that
here modern observers are 'uninvited guests in a system of communication designed for others'. But, for the set of hand stencils at
Cosquer, it does seem possible to go further: to say that the five
hand positions that were used meant something, and that the
twenty-seven others were left out because in this system they had
no meaning. And it is almost impossible to avoid the speculation
that the hand positions recorded had something to do with
counting. They form a sequence consisting of extending the digits
one after another; they are one of the easiest ways you can count
to five on your fingers. They are a good candidate for the earliest
counting vocabulary recorded: a vocabulary not of spoken words
but of gestures; a sign language.

Other caves with hand stencils display different vocabularies,
using finger combinations that are not seen at Cosquer. Many
locations show complete hands only, or record a more restricted
set of folded-finger gestures. Perhaps the language of hand gestures
differed from place to place; perhaps chance survival of fragile
images in difficult environments has played a role. That does not
detract from the consistency and suggestiveness – to say no more
– of the evidence at Cosquer.

The gestured numbers, if that is what they are, do not all appear
with the same frequency. The complete hand is by far the most
common gesture recorded; those showing smaller numbers of

fingers are progressively less common, with the interesting exception that four fingers is less common than three. A sign for zero (logically, you would fold down all five fingers and stencil just your fist) does not appear at all. The signs never seem to appear in relationship to each other – sequences of numbers, pairs – nor do they occur in two-handed groups that might suggest counting to numbers beyond five. If this was a number vocabulary, in other words, it ran *only* from one to five.

As with other prehistoric counts, it is impossible to be sure what was being counted. But the proximity of images of animals, including animals with signs on them such as barbs that might be schematised weapons, means that a strong possibility is a count of animals taken. Another would be a count of visits to the cave.

Nor is it certain why people chose to record these gestures by stencilling in this particular location. To turn a gesture into a stencil is to extend it beyond your body and – potentially – beyond your own lifetime, to transmit it to people you have not met and to generations not yet born. Yet to place the stencil on the wall of a deep cave requires those who wish to see it to visit that location. Like certain tallies, these counts were not portable, could play no part in exchange with other groups in other locations. Their position in a dark cave at the end of a long tunnel hints at a special purpose, but anything from a game or a joke to an important ritual is equally possible.

It is not clear how long this use of Cosquer went on: years, centuries, or even longer. But eventually it ended and the cave, it seems, was neglected for millennia. It was then used again, around 19,000 years ago. The walls began to become a palimpsest, made by people who knew nothing of each other beyond what was recorded on the walls of the cave. This later period saw some of the hand stencils destroyed, scratched or overwritten with new signs, changing their meaning in ways that are just as impossible to recover as is the original significance of the cave and its decorations. A wealth of paintings in black were added at this stage: over 150 images of horses, bison, deer as well as seals and – not yet extinct – great auk.

The cave was abandoned once again, and as the ice began to retreat 16,000 years ago, the sea rose towards its entrance. Some time around 5000 BCE the cave mouth was closed by water, along with those of many other caves along the coast of what is now France, in the area of Marseilles and Cassis. Others, surely, had been used by prehistoric people. By about 500 CE, the sea had reached something like its present level, with the cave mouth more than 30 metres below the surface. Sea water destroyed perhaps three-quarters of the paintings, stencils and engravings, and the cave remained unknown until its rediscovery by a professional diver in the 1980s. Its art has been documented since 1991; it is the only painted cave known today whose present entrance is under sea level. It is also the only location with evidence of prehistoric counting in this particular form: an astonishingly lucky survival. Without it, the finger-counting of the Stone Age would be gone for ever.

Counting words

A final form in which counting may have existed in the African Stone Age is in words. Spoken words leave no trace of themselves, of course, and it would be many millennia before people devised ways to record speech in writing. There is no amazing survival like the cave at Cosquer to report what Ice Age voices sounded like. The clues are plentiful, as are the theories: but certainty is not to be expected. The mystery is once again part of the story.

Language arose in the human species almost certainly in Africa, after the split from the chimpanzee lineage (six or seven million years ago) and before the dispersal to populate the rest of the world (seventy-odd thousand years ago). Within that very broad range, almost every possible hypothesis has been tried, from an early or gradual origin and a period when language was shared among several hominid species to a late and/or sudden appearance, part of a package of modern behaviours whose most obvious result was the exodus from Africa itself. More gradualist scenarios seem to be in favour at the moment, and genetic evidence taken from certain relatively isolated African groups speaking what seem to be languages of very ancient lineage might seem to support that, providing a minimum date for the origin of language of perhaps 130,000 years ago. Much effort has been made to argue that certain developments visible in the archaeological evidence – such as more complex tools – required language for their development or transmission, or that symbolic behaviours such as the use of ochre as a pigment or the manufacture of beads and the marking of surfaces – as at Blombos – are functionally equivalent to language; but these claims have been denied with equal force, and it seems fair to say that there is no consensus as yet. Claims for a 'language gene' or set of genes have proved inconclusive, and if a change in human anatomy was involved – such as subtle change to the way the brain was organised – the archaeological remains cannot confirm it.

Whenever language did arise, there may well have been times and places when humans had words for certain things, but no counting words. Counting words and verbal counting routines are not, in fact, absolutely ubiquitous among cultures living today. Despite the fact that virtually all modern humans are capable of acquiring counting words if they are taught them, not all in fact do so, and not all take the sequence beyond a few words or attach any great importance to it. The value placed on counting varies from community to community, and it does not correlate with, and should not be identified with, some such abstraction as 'sophistication' or 'civilisation'. It is quite unimportant to some. Thus there is no positive reason to assume that counting words were part of the 'ground floor' of human vocabulary; nor that they must have appeared at any particular stage in the development of language. It is quite possible, indeed, that beads, tallies and even numerical hand gestures were made by people who had no way of counting in words.

On the other hand, nearly all of the roughly 7,000 languages spoken today – or extinct recently enough to be documented – do contain at least some counting words: some terms that can give an answer to 'how many' that is more precise than 'few', 'some' or 'lots'. It has seemed to many a reasonable speculation that at least some of the people of the African Stone Age therefore had verbal counting routines alongside – or instead of – their other ways of counting.

Perhaps, then, certain human groups possessed ritualised sets of sounds that they made to accompany repeated actions, and over time those sounds acquired something of the character of a 'wild' number line. Perhaps those sounds eventually came to be thought of as words just like other words: came to be used as adjectives describing the *how many* property of groups of things or sequences of actions. Perhaps other sets of words were then derived from them to answer related questions like *how many times* or *which in a sequence*. (The range of ways the list of counting words may be further modified is longer than you might think. In English there are not just cardinal numbers *one, two, three* and ordinals *first, second, third* but also sequences like *once, twice, thrice*; or *single, double, triple*; or *single, twofold, threefold*. Other languages have similar ranges of options, known by names such as the 'distributive' and the 'frequentative' numerals: they are used to answer questions like *how many each* or *into how many parts*. They are widespread enough among documented languages to suggest that some selection of these types of number word may have occurred in very ancient languages.)

There is no way to know what these ancient number words sounded like. Attempts to reconstruct vocabulary and phonology from several tens of thousands of years ago are occasionally made, but they have not found acceptance by mainstream linguists, and for good reason. It took just a thousand years for Latin to turn into Italian (and Romansh, French, Catalan, Spanish, Portuguese . . .). Multiply that degree of change – in sounds, structures and meanings – by seventy or more, and it is clear that the noise will overwhelm the signal. It is quite impossible for precise information about the sound or structure of language to survive from the time before humans left Africa.

Perhaps some Stone Age humans went further in verbal counting. If your set of counting words has just five or seven or ten members, say, it is quite acceptable for them to be arbitrary, unrelated sounds. But beyond a certain point, in order not to make intolerable demands on people's memories, in order to be learned by children and retained by adults, there will be a need for some sort of internal structure. The norm today is to reuse a fairly small set of words over and over, employing multiplication and addition to reach higher numbers. ('Seventeen' meaning *seven and ten*, for instance; 'twenty-four' meaning *two tens and four*.) A number that regularly occurs as a multiplier in such a system is called a 'base'; it is perfectly possible, and even quite common, for a system of number words to have more than one base. Again, this feature is not ubiquitous, but it is common enough in modern languages for it to be plausible that some of those who counted in Stone Age Africa may have used counting words with this kind of internal structure.

Once again, it is not possible to know what number bases were used in the distant past. Ten is by far the most popular base in the world today, and it is tempting to suppose a cause-and-effect relationship with human anatomy: the ten fingers of both hands. It has been many times pointed out that in a lot of languages the word for five derives from the word for hand, or that the word for digit also means number. But it is not clear that creatures with five digits on each of their four limbs must always arrive at 10 as a base for counting. It is perfectly possible to arrive at 4 as a base, by counting fingers but not the thumb, or to arrive at 6 by incorporating the wrist; or indeed to arrive at a base-5 system by using just one hand. Counting toes as well will give you base twenty, and it is equally possible to arrive at base 12 by using the thumb to count the three joints of each of the four fingers . . . History and ethnography in fact show a wide selection of other bases (2, 3, 4, 5, 6, 7, 12, 15, 20 . . .) being used at various times and places. The prevalence of 10 today is in part the effect of happenstance, the result of its adoption by a small number of successful languages and language families. There is, in short, no good reason to project onto the distant past today's widespread preference for tens.

A cautious hypothesis, then, is that the hunter-gatherer groups of the African Stone Age had just as much variation in their sets of counting words as the cultures to be observed today. That is, some had none, some had short lists of counting words with little or no internal structure, and some had larger sets of counting words structured with number bases. (Many may have had the feature called 'grammatical number': a distinction between singular and plural forms for certain kinds of word; or even between a larger set of options such as singular, dual (two items), trial (three) and plural. But not all need have had this feature, since not absolutely all living languages do.) The different systems were transmitted culturally; they were not biologically determined. And they must surely have changed over time, perhaps developing out of all recognition over hundreds of generations.

So the Stone Age human environment contained – at certain times and places – at least three sets of physical objects that could have been 'wild' number lines, that could have supported a transition from not-counting to counting: fingers, beads on strings, and marks on surfaces. It is possible that a conventional series of sounds accompanying repeated actions might have served a similar function and become the first counting words. It is also possible that unbound counters – which would be impossible to distinguish from ordinary pebbles or fragments of bone or other material in the archaeological record – also supported counting in the Stone Age world, although since they do not possess a natural order their ability to function as a 'wild' number line is less clear.

How did these different precursors to counting – and these different early ways of counting – interact? Which came first? Many scenarios have been proposed, but there can be no clear dénouement to the mystery story that is the origin of counting.

Some have argued that language must come first: that, for instance, no one would make tally marks without having words for the numbers tallied. Or that there may have been words for the subitisable numbers from one to four at a very early stage, before counting was later extended by fingers, beads and tallies. Others – probably representing the balance of opinion at the moment – believe that language is not essential for counting: that you don't invent words for things you have not yet experienced, and that fingers, beads and tallies more likely came before counting words. Or perhaps that counting words evolved alongside material ways of counting.

But perhaps the search for a single scenario is itself a mistake. In the context of prehistoric art, it has been well said that development should be imagined not like a line or a ladder but like a bush. In the same way, perhaps the development of counting in the Stone Age was more like a knot garden than a footpath. More like a collection of short stories than a novel. The scales of time and distance are enormous, even taking the most conservative estimates for the rise and spread of *Homo sapiens*, and it is perfectly possible that, over thousands of kilometres and tens of thousands of years, different ways of counting arose and were forgotten – even multiple times – by people who never met each other, in Africa and beyond.

Whatever the truth of these matters, ways of counting with beads, tallies, fingers and words are found the world over, and it is all but certain that by the time modern humans began to populate the world beyond Africa – perhaps 70,000 years ago – they were doing all of those things. That the stage was set for the immense, global phase of the human history of counting.

The numbers

At some point in the history of counting, people started to treat counting words not just as noises-in-a-sequence, but as properties of collections of objects: as adjectives. And at some point they started to treat them not only as adjectives, but sometimes as nouns (in some languages, there are concomitant changes to the exact form of the words). That is, words like 'three' would now denote not only a certain property that collections of events or objects could have ('I have three axes'); they would also denote certain abstract objects ('three is a small number'). Not all languages do this, and in many the number words occupy an uneasy position between adjective and noun, with some – but not others – capable of being modified for case or gender, or even possessing separate plural forms (in English you can speak of 'hundreds' of something but not really of 'ninety-nines'). But most – not all – cultures do appear to have a category of 'numbers', whose names at least are built from the counting words.

To study the relationships between numbers is called arithmetic or, in its higher reaches, number theory. To place them in correspondence with other things or ideas – beyond using them as a set of mere labels – is called numerology. Such studies almost certainly stretch back into prehistory, and they appear very early on in written texts; there can be no knowing when or where exactly they started. Some would certainly see artefacts like the tallies from Ishango or Laussel as reflecting early attempts to study number rather than merely to count.

But what are numbers?

A common intuition is that behind different ways of counting there must lie some *thing* which guarantees their integrity and their mutual consistency: some set of abstract objects with the right properties to account for everyone's experiences of counting. The numbers, in other words: with a reality independent of human minds; indeed, independent of physical reality altogether, living eternally in a separate realm. In this view, the nature of things constrains the ways people can count, and the history of counting might be seen as one of striving towards the real truth about numbers. The numbers and their properties would then be another of the roots of human counting: surely the most important root, in fact, underpinning both the innate perceptions people can have of number and quantity, and the properties they can find in sets of beads, tallies, fingers and counting words.

That intuition is not universal, however, and it resists satisfactory proof or disproof. If the numbers live outside time and space, just how do they 'get in' and affect the world or become perceptible to human minds? And is it really credible that not just 1, 2 and 3 but 2,938,239,856,837,641 existed from the beginning of time waiting for someone to have a use for them? Surely there must be a less extravagant way to account for what happens when people count.

Others would say that numbers have no reality except as the structures of things in the world: that numbers are abstractions from experiences that people have, and that their consistency over time and space is provided by the consistency of those experiences. People's intuitions about small numbers, in this view, are memories of experiences with small collections of objects, while their shared ideas about larger numbers are the consequence of all playing the same arithmetical games by the same rules. There is no need to suppose that 2, 3, or any other number has an existence separate from the properties of collections of objects or sequences of events. Thus, if there are constraints on human counting, they are the ones provided by the accidents of human physiology and by the kinds of environment in which human beings have lived over the millennia.

Others, again, would say that whatever the truth about the real nature of numbers, human brains are, as it happens, hardwired to view numbers in a particular way: to recognise certain kinds of structures in experiences, or even to impose those structures on them. In this view, at least some of the constraints on the ways human beings count arise from the accidents of how their brains are organised; which are in the end the consequences of what was adaptive in the evolutionary past, not necessarily of what was true. This view opens up the rather staggering possibility that the numbers are to some degree a shared illusion.

There is no consensus about such matters. Really convincing evidence for or against any of these positions has proved to be extremely elusive. In the ancient Greek philosophical tradition, Plato was on one side (numbers are real), Aristotle on another (numbers are abstractions from experience). Among their successors, no later philosopher has come up with an argument that convinces *everyone* of one position or the other. Thus, it remains a vexed question how far human counting is constrained by the givens of human physiology, how far it is constrained – if at all – by the nature of things, and how far it is freely chosen and transmitted as a matter of human culture.

Luckily, for the purpose of exploring the many ways human beings have in fact counted, there seems no need to choose between these positions. This book will try to keep an open mind.

PART 2

Branches

3

Counting with words and symbols in the Fertile Crescent

Walk out of Africa, following the fertile land. It seems that always meant the Nile valley – the only way to cross the Sahara – and then the hospitable strip of the Levant between the Arabian Desert and the Mediterranean. Then cut through the Anatolian mountains and, turning right, follow the Tigris or the Euphrates, perhaps all the way down to the sea: the modern Persian Gulf. The route was important for humans, and for animals and plants, expanding out of Africa over tens of thousands of years. The whole route – now sometimes known as the Fertile Crescent – became an important site for human culture. It lies between three continents – Africa, Asia, Europe – and indeed between three tectonic plates, whose meeting and movement created the great valleys and mountains of the region: the Dead Sea Rift, the Tigris and Euphrates valleys, the Anatolian and Caucasus mountains. It is a region, consequently, of enormous variation: marshes, deserts, slowly shifting river valleys and mountains.

One branch of the story of counting flourished in the Fertile Crescent, from the beginning of writing around 5,000 years ago, through to the Hellenistic period more than three millennia later. In Sumer, Assyria and Egypt, from the earliest writing to the final flourishing of the Egyptian script under Greek domination, and from the world's first city to the last centuries of the Pharaohs, counting was always present, used for administration, for boasting, for keeping track, for proving who owned what and who had paid whom.

The region was of course a patchwork of different cultures speaking different languages and doing things – including counting – in different ways. But it is fair to identify words and symbols as consistently the important features of counting in this place and time. This is a place where something more specific can be learned about number words: the oldest number words for which there is secure documentation come from Sumer, in Mesopotamia. And this was also one of the places where a crucial new item was added to the human repertoire of ways of counting: number symbols.

Sumer: Counting symbols

Squat on the ground in the sunlight, in the greatest city in the world. Shape a lump of clay into a flat tablet, to fit in the palm of your hand. Take a reed, scratch a drawing in the clay: a goat. Turn the reed on end and press holes into the clay: 'dish, min, esh'. It means 'three goats'. Five thousand years later, it will still mean 'three goats'. The bright light of writing changes everything.

The eastern end of the Fertile Crescent, now sometimes called Mesopotamia, contains in itself much of the region's diversity. Near the rivers, annual flooding makes the land fertile. Marshes skirt the coast. There are resources ranging from fish, clay and reeds, wild barley and wheat to wild sheep, goats, cattle, pigs and camels. The potential for hunting and fishing is great. People have been in the region for many millennia.

As the climate improved from its ice age low, this was one of the regions in the world – the first – where humans took control of their food supply by domesticating cereals: selecting, sowing, harvesting, storing. Settlement moved from northern Mesopotamia to the plains beside the rivers; by about 7000 BCE, villages dependent on agriculture existed throughout the Near East, wherever there was enough rainfall to sustain it, and the aeons-long dependence on hunting and gathering faded.

By late in the fourth millennium BCE, the plains in the south of what is now Iraq were becoming densely populated, supported by fertile soil and increasingly elaborate irrigation. Settlements became larger, and the first cities arose.

To its inhabitants, the greatest of these was simply 'the city': '*Uruk*'. It lay in the very south of Mesopotamia, on the Euphrates, near the Persian Gulf. It may have formed from two settlements on the two sides of the river, and by 3000 BCE or so it covered 600 acres or more; two and a half square kilometres. The population may have approached 40,000. The world had, literally, never seen anything like it.

Physically, it was dominated by a temple complex covering 75 acres: ceremonial areas set on platforms, some built of stone, their walls decorated with mosaics of white, black and red clay and visible for long distances over the plain. Culturally, the city dominated its surrounding villages, consuming much of the food they produced and setting up a specialisation of labour among producers: fishers, farmers, gardeners, hunters, herders; but also weavers, potters, metalworkers and other kinds of specialist. Timber, stone and metals all had to be imported, and long-distance networks of barter existed as a result.

———

All of this required organisation in various senses. There was a ruler and some sort of city assembly, though the working of these institutions in the fourth millennium are very obscure. To organise people and things on this scale required language, of course, and the inhabitants of Uruk and its area spoke a language called Sumerian, which has no known relatives and whose grammar and even set of sounds are not completely understood.

And it required counting.

Sumerian had a set of number words: *dish* or *ash, min, esh, leemu, ya*. This primary set from one to five was compounded to form words for six through nine: *ya-ash* becoming *yash*; *ya-min* becoming *imin*, and so on. Beyond that, Sumerians counted in a decimal system, with ten being *hu*, twenty *nish*, thirty *ushu*, and so on up to sixty: *gi* or *esh*. The limited evidence suggests that for larger numbers, 60 was used as a base, so that you would have said the equivalent of 'sixty, ten, three' for 73 and 'two-sixty, forty, two' for 162. *Shar* appears to have meant 3,600 (that is, 60 times 60) and (some scholars say) *shar-gal* 216,000 (that's 60 times 60 times 60).

Indirectly, the use of five and ten in the set of counting words suggests an interest in the fingers used as a set of five, as in some – but by no means all – other languages and language families. It is very likely that Sumerians counted on their fingers at times, but there is no direct evidence of the fact: no hand prints like those from Cosquer. Equally, it is possible that they sometimes made tally marks, but none have been found from early contexts.

More probable is that counters were used in the Mesopotamian world. Unmarked pebbles cannot be confidently identified as counters in the archaeological record, but from as early as 8000 BCE, small clay objects appear at sites in Syria and Iran, and one attractive interpretation is that at least some of them, at least some of the time, functioned as counters. Thousands have been found all over the Near East from the Mediterranean to Persia, and it seems very likely that those with more elaborate shapes or those pierced for – perhaps – stringing, had other uses and meanings besides number (these become more common after about 4000 BCE). Yet, as with beads from much earlier periods, they witness at least to the presence in the human environment of objects whose manipulation could form an ordered sequence and come to be used to count. Exactly when and how frequently that happened in this case seems likely to remain obscure.

By the late fourth millennium, the demands of running long-distance trade and managing the supply of food and materials to the world's largest city prompted a variety of experiments in information storage. Wet clay could be used as a seal, much as wax was at other times and places – say around the neck of a sack or the cover of a jar – and could bear the impression of a stamp for extra security. Simple stamping had already appeared by 6000 BCE. More sophisticated than a simple flat stamp was the cylindrical seal, which was rolled across the wet clay to provide an impression of its design. These first appeared in the mid-fourth millennium; they could be made of shell, bone or stone and they were small – rarely more than 3cm long – making their carving a matter of considerable skill and time, since the scenes carved on them were sometimes elaborate and detailed. Presumably the owner of the seal could be identified from the design by those who needed to know. The increasing use of seals speaks of a society in which memory, word of mouth, and trust were no longer adequate tools to identify and protect property, an effect presumably of that society's size and complexity.

A different experiment in record-keeping also used clay, but in the form of hollow spheres or envelopes, into which small clay tokens could be placed before the envelope was closed. Very possibly the size, shape and number of the enclosed tokens represented types and numbers of real-world objects, and it is possible that this technique related directly to the use of tokens to signify the same things in other contexts. The outside of the envelope could be marked with a seal or with symbols corresponding to the number of tokens inside: only a handful of examples are known, and it is not clear how common this was or exactly how it worked. It is, in fact, the only direct evidence for clay tokens denoting numbers in the region. These clay envelopes were in use for perhaps a millennium, from about 3700 BCE on.

All of these clay-based recording systems still relied on people's memories to some degree; their capacity for recording details was limited, as was their ability to prevent tampering. It is tempting to arrange them into a developmental sequence from the more simple to the more sophisticated, but the archaeological contexts – sometimes far apart – in which they are found do not provide enough chronological detail to confirm such a story. Perhaps the better reading is that the later fourth millennium, and particularly its final two centuries, were a period of innovation and experiment, at various places and in various directions.

Number symbols from Uruk.

Yet another possibility was to take a free-standing flat tablet of clay and mark it with impressions denoting the things to be recorded: the impression of a cylinder seal or a stamp seal; a quick stylised sketch of an object (a goat, a sheep) done with the sharp end of a piece of reed; an abstract sign functioning as your personal 'mark'; a few quick stabs with the thick end of the reed to count how many (goats, or whatever). Left in the sun, the clay would dry and become a durable record of the information it contained. It could be moistened again and modified, though tampering with the text in this way would surely be fairly easy to detect; the whole tablet could be dunked in water prior to erasure and reuse; or it could just be thrown away once its purpose had been served. Artefacts like these began to be made perhaps as early as 3500 BCE, and the simpler ones contained just a single unit of information: say one or more stylised signs plus a set of reed-stabs denoting a number. At first, the number marks were simple tallies; one example from a Syrian site has twenty-two marks with no sign of any internal grouping.

By about 3200–3100 BCE, this method had become distinctly more elaborate and sophisticated. The flat tablet of clay remained (typical sizes were from perhaps 5 to 15cm across, but both larger and smaller were possible), as did the reed stylus. So did the convention of scratching signs with the thin end of the reed and impressing number notations with the thick end. The number of signs had burgeoned to several hundred – nearly a thousand in some estimates – and the origin of some of them as pictograms had become obscured; a high proportion were now to all appearances wholly abstract, their meaning needing to be learned and handed down from person to person.

Furthermore, the surface of the clay was now divided up with lines into boxes, each holding a separate unit of information. The order of signs within each box still seems to have been quite arbitrary, but the arrangement of boxes on the tablet was itself perhaps starting to be standardised, the relationship between one box – one unit of information – and another a matter of conventions understood by those who used the clay tablets.

The number signs themselves had also undergone some development by now, and had acquired something of the same complexity as the non-numerical signs. To count discrete objects, the system mimicked the structure of the spoken Sumerian number words, with symbols for 1, 10, 60, 600, 3,600 and 36,000. For 1, the symbol was a small triangle; for 60 a larger triangle; and for 3,600 a large circle. The symbol for 10 was a small circle, and that symbol could be added inside the signs for 60 or 3,600 to yield 600 or 36,000. Each sign was repeated as many times as needed, and signs for larger numbers conventionally came to the left, smaller to the right.

To count particular commodities such as grain, land, beer or sheep, though, there were different systems, with not only different sets of signs but different relationships between them. To count processed grain, for instance, there were symbols for ½, 1, 10, 60, 120, 1,200 and 7,200; for areas the symbols stood for 1, 10, 60, 180, 1,800 and 10,800. Presumably these were based on conventional measuring instruments such as rods and bowls and their relationships, the written symbols reflecting the fact that, say, there were ten small bowls in a large bowl, and so on.

In all, there were at least twelve different systems of number signs, perhaps as many as fifteen. Some symbols were specific to one system, but others were reused across more than one, making them ambiguous unless the reader paid close attention to the context. A small round impression could stand for ten discrete objects, but also for eighteen large units of land, thirty small units of grain . . .

———

The marks on the tablets from Uruk were probably the world's first writing. If there were any earlier attempts to represent human language through marks on surfaces, they have not survived, or

archaeologists have not yet found them. Thus, from Uruk's period
of apparently concurrent, competing ways to record information
and signify the world more durably than speech and more reliably
than memory, this was the innovation that ended up being most
long-lived and important; although clay tokens, seals and stamps
would also continue in use for centuries more. Admittedly, this
very earliest writing performed only some of the functions that
would later be associated with written scripts. Neither the order
of the words nor their specific grammatical form was represented,
for instance. The signs, indeed, were not an attempt to record the
spoken language completely; the tablets should be compared to
invoices and receipts, not to pages of prose.

Equally, the numerical marks from Uruk are the world's first
number symbols: more complex and abstract than one-to-one
tallying, and clearly deriving their structure from the number words
that preceded them. Using Sumerian words, you said the number
of tens then the number of units; or the number of sixties,
then of tens, then of units. In Uruk's symbols you did much the
same, making a mark for each sixty, each ten, and each unit. When
speaking in Sumerian of, say, amounts of grain, you reported the
number of large bowls followed by the number of small bowls; in
symbols you made a mark for each large bowl then a mark for
each small bowl. So the symbols followed, apparently quite closely,
the way quantity was talked about in Sumerian.

Around 5,000 tablets have been recovered from Uruk, where
after their useful life was over they were repurposed as building
rubble in the temple complex. They apparently date from soon
after the system of symbols was invented. They were used within
what was evidently the complex domestic economy of the city, in
which it was of real importance to know how many units of
different commodities had been transferred from owner to owner
or from location to location. Clay is not a very flexible medium,
and its limitations shaped how the process could work: it would

harden in the time it took to move a sheep from one side of the city to the other, say, so there could be no running accounts built up over days or weeks. Indeed, if a set of tally marks or a set of raised fingers relates to the activity of counting rather as smoke does to fire, more structured notations leave a wider gap between the act of counting and the trace it leaves behind. First you count (or indeed calculate), then you write a notation that records the result. There was no question, then, of a scribe traversing the city with a set of wet clay tablets in a satchel, witnessing transactions and writing them down in real time. It is likely that most records were made after the fact and that there was an element of unreality about them: that their purpose was to establish responsibility for certain assets rather than necessarily to report movements of goods and people that had literally happened.

That said, the clay tablets speak of a concern to illuminate even the smallest details of the city's economic life: information about animals, barley, people and land was reported with painstaking accuracy. They speak of a culture in which counting (and measuring) had become a critically important part of the economy. Some were later summarised into secondary accounts reporting totals and subtotals, say of temple offerings over a period of years. It is possible that the information was used – was intended to be used – for planning, such as the apportioning of rations or of labour.

Who did the writing, and the reading, and the planning? The number of recovered clay tablets is enormous, but marking clay takes little time and it would not have occupied more than quite a small number of people on a full-time basis. The Uruk term for them was *umbisag*, denoting an accountant or scribe: the world's first numerate professionals. It was a new role, a person who neither made goods nor wielded political power but who stood between the two, managing the production and movement of goods on behalf of the city's rulers. They were few, and they had skills and therefore power that the vast majority – including quite probably the elite rulers themselves – did not. In the seventh century BCE, nearly two and a half thousand years after the invention of writing, an Assyrian king could still boast of being able to read and write as though that was unusual. As historian Marc Van De Mieroop puts it, 'people must have realized how important documents were, how they protected them against accusations of theft or careless loss. The scribe had thus certain esoteric powers, a control over people's lives that was expressed in ways most of them could not control.'

———

Even during its first centuries, the Sumerian system of marks on clay was changing and developing. Perhaps as soon as there was a community committed to sustaining this new way of recording information, that community set about, both by accident and by design, to refine the system. It became conventional to arrange the signs in lines read from left to right and to place them in the order in which they would be spoken; the lines themselves were arranged in columns whose number depended on the size of the tablet. It was soon realised that impressing the stylus into the clay was quicker and more attractive than scratching lines, and so the older curved lines were replaced by straight ones. The scribe would now press the head of the stylus into the clay and then push it down to the side, producing a characteristic wedge-shaped mark. The signs, already mostly stylised, became completely so, and consisted of one or more wedges in various configurations. (It could now properly be called a *cuneiform*, 'wedge-shaped', script.)

The number signs developed in the same way, so that numbers from two to nine were now shown by a single wedge repeated the appropriate number of times; the number one by a slightly more elaborate single *T* shape, and ten by a *one*-sign rotated to the left.

The structure of the system remained the same, though, and a range of different systems for counting and measuring different things continued in use. There was some reform during the course of the third millennium. Gradually the systems seem to have been reduced to five: one for lengths, one for areas, volumes and bricks, one for liquid capacities, one for weights, and one for discrete objects. They continued to have various bases, in which multiples of sixty were important but not all-pervasive. Lengths, for instance, had thirty fingers in a cubit, six cubits in a reed and two reeds in a rod.

Perhaps more important than these changes in the appearance of the script were moves towards recording a larger proportion of the words actually spoken. Many Sumerian words had only a single syllable, which meant that the sign for a word could also be used to represent its sound. That created the possibility of spelling out other or longer words, including for instance people's names, on a phonetic basis. Thus the sign for 'reed', sounding '*gi*', was reused for the word 'to return', which happened to sound the same. A single sign could end up having as many as twenty different meanings, and you had to know the language well to interpret the script.

The system of receipts and summaries continued in use, and up to 90 per cent of the documents excavated from the third millennium are of that quantitative type. The remaining 10 per cent include an increasing proportion of Mesopotamian literature: the world's first written myths, set in the world of Gilgamesh and Utnapishtim. They also contain lists of words, compiled by and for trainee scribes. The longest contain thousands of words, and give a vivid sense of the difficulty of becoming proficient in the system of cuneiform writing.

Techniques of calculation, similarly, had to be learned and practised at the cost of considerable effort. Some clay tablets were used for artificial calculation exercises, such as invented accounts, or problems like how many workers a given quantity of grain could feed or how much land was represented by a given survey.

Addition is fairly easy with a system like the Sumerian ones; you merely need to write down the complete set of signs represented by the two quantities to be summed, and if any sign occurs more times than is allowed you do some replacement, setting a 10 in place of ten ones, a 6 in place of six ones, depending on the system being used. Subtraction is not much harder. But multiplication and division require some work, as indeed they do with any set of number symbols. For all the calculating that was involved in producing these accounts, traces of written calculation – rather than its end results – are very rare indeed in the early cuneiform record, and it is likely that a non-written technology was used to calculate: an abacus with beads on strings, perhaps, or a board on which counters moved. None has survived; neither has any depiction of one.

———

Finally, the later third millennium saw an intriguing innovation in the cuneiform way of writing numbers. Scribes began to simplify their calculations and the conversions between different systems, by taking quantities expressed in the traditional systems and rewriting them all in the system normally reserved for counting discrete objects: the one that went 1 – 10 – 60 – 600 – 3,600 – 36,000. The convention that the signs were written in descending order of size was now long established, and the scribes further simplified the notation by abandoning the traditional range of differently shaped signs for the larger values; instead they just used the 'one' sign – an upright wedge – for 1, 60 and 3,600, and the 'ten' sign – a slanted wedge – for 10, 600 and 36,000, relying on the order in which the signs appeared to convey which number was intended. You would now read \\||\\\\|||| as two 600s, two 60s, three 10s and four 1s (equal to 1,354).

These were understandable simplifications motivated by practical needs. Scribes, having written a quantity in the new notation, could calculate with it efficiently, and then convert the result back into whatever traditional system was appropriate; this new notation was not used for record-keeping. Some surviving clay tablets include tables specially for multiplication and division with this 60-based system, namely tables of number pairs whose product is sixty.

The earliest datable calculations in the new notation are from 2039 BCE. A system that relied on position rather than symbol shape to denote numbers in this way was a flexible tool; it was, interestingly, an indefinitely extensible system, with no particular upper limit on the size of the numbers that could be expressed.

Over the following millennium, it became usual for scribes to turn everything into the new positional notation when calculating. By the first millennium, it was in use in Babylonian astronomy, and its emphasis on the number 60 is the ultimate ancestor of the division of circles into 360 degrees, and hours into sixty minutes.

A question that arises naturally to a modern mind, and must have occurred fairly early to users of this notation, is what to do if there is, so to speak, a blank column. How do you express the number made of two sixties and three units but no intervening tens, for instance? Some sort of sign meaning 'blank', or just a blank gap on the clay, seems to be needed. The earliest surviving example of a solution to the problem is in a table of squares from Kish, dating to the seventh century BCE: it has a placeholder identical to the sign for '30', performing the 'blank' function. Signs with this function would become common in later Babylonian arithmetical and astronomical number writing.

Cuneiform number symbols had come a long way from their origins, both in their appearance and their functions. Elsewhere in the Fertile Crescent, they were being put to very different uses from administrative record-keeping, or indeed astronomical calculation: they had become part of the tradition of royal self-advertisement, display and boasting.

Tiglath-Pileser I: Counting plunder

The city of Katmuhu, near Assyria; early in the eleventh century BCE. Two scribes survey the field of battle, the killed, the captured and the plunder. One uses a stylus to keep track, counting under his breath. Later the record will be written up and become part of the royal annals:

'I captured in battle their king, Kili-Teshub, son of Kali-Teshub, who is called Errupi. I carried off his wives, his natural sons, his clan, 180 copper kettles, five bronze bath-tubs, together with their gods, gold and silver, the best of their property.'

As well as undergoing a range of developments in its original Mesopotamian location, cuneiform writing spread throughout the Near East. The system was exported early to other cities in Iraq and Syria; the evidence is fragmentary, but numerical symbols like those from Uruk have been found at a number of sites in the late fourth millennium, and it is possible that the original invention was a more widely distributed affair than the evidence now shows.

The later changes made cuneiform a flexible tool, capable for instance of being used for literary texts and letters. They made it capable of being used to notate languages other than the original

Sumerian. Cuneiform would eventually be used to record about fifteen different languages across the Near East; at its peak, it was used from Turkey to Egypt and from Lebanon to Iran. In fact, cuneiform on clay would remain in use for over 3,000 years, making it the longest-lived writing technology the world has yet seen, its history longer than that which separates the present from the fall of Troy. The Uruk system of record-keeping was the basis for administrative systems over the whole of that period.

One of the languages most associated with cuneiform is Akkadian, a member of the Semitic language family, whose members also include Hebrew, Aramaic and Arabic. For reasons that are now very unclear, it replaced the unrelated Sumerian as a spoken everyday language in Uruk during the third millennium BCE. It went on to become a widely used language in the Near East, functioning as a common tongue for purposes such as diplomatic correspondence. In the dialect of Babylon, it was a major language of literature over the whole region; in the dialect of Assur it was the language of an important city and, at times, an empire.

Assur stood on the west bank of the Tigris, about 100 kilometres south of modern Mosul. It overlooked the river to the east from a rocky promontory. This area had a different climate from Sumer, with more rainfall, making agriculture easier and less dependent on irrigation. Stone was available locally and did not need to be imported; equally, timber and metal could be found close by. Conversely, the marshes and their characteristic products were absent. Assyrian merchants operated as far away as central Anatolia.

Ruled by a king and an assembly, Assur functioned a little like Uruk, with agricultural hinterlands providing cereal and animal foods and the city housing specialised crafts. Assur was established as a city by the middle of the third millennium, and it had periods of relative power and periods of collapse. From the mid-fourteenth century BCE, a series of ambitious kings made it an important political force in the region. Both diplomacy and military conquest played a part, with Tukulti-Ninurta I in the late thirteenth century – for instance – successfully attacking Babylon itself. He sacked the city, trod on the neck of its king 'as though it were a footstool' and took him in chains back to Assyria. There was campaigning in all directions at various times. On the diplomatic side, Assyrian power was eventually recognised by its regional rivals the Hittites, by the Babylonians, and even in Egypt; Assur was acknowledged to control the Tigris valley and the plains to the east, as far as the Taurus mountains to the north.

———

As at Uruk, resources were controlled through cuneiform accounting, which enabled them to be mobilised for building, irrigation and agricultural schemes. Like Sumerian before it, Akkadian had number words (*isten, sina, salasat, erbett, xamsat* . . .). Like Sumerian, it used a decimal system for smaller numbers and showed at least traces of a base-60 system for larger ones, with special words reported for 60 (*sus*) and 3,600 (*sar*) but also for 100 (*meat*) and 1,000 (*lim*).

To write numbers down, Assyrians used a hybrid system. For 1 and 10 they took over the symbols from the Sumerian positional system: a vertical wedge and a slanted one respectively, each repeated up to nine times as necessary. Beyond that, they used ideograms to denote 60, 100, 600 and 3,600. Over time, the 60s, 600s and 3,600s fell out of use and it became more usual to use a purely decimal system with just the signs for 10, 100 and 1,000. But a fully positional system was not consistently used.

They used their written numbers for accounting, for institutional record-keeping, much as the Sumerians had, although there were differences in detail. Akkadian-language accounting – even at Uruk itself – tended to be more interested in estimates and predictions with standardised round numbers than its early Sumerian equivalents;

more interested, perhaps, in overview than in detail. The two styles could coexist in the same archive: 'approximation tended to be used for calculating the assets and rights of the wealthy and the institutionally powerful; precision for the dues of the dependents.' At Assur, numeracy also became a less exclusive property than at Uruk; it was used by merchants who trained less formally and depended less heavily on prestigious Babylonian models. And written numbers were also used by the Assyrian kings as part of their programme of public self-advertisement and propaganda: a quite different function for counting than any to be seen before.

From an early date, kings had had inscriptions carved onto stone monuments. From the time of Tiglath-Pileser I (1114–1076) they burgeoned into a whole new genre of self-glorification: the royal annals, remarkably detailed chronological accounts of royal and military actions. The first surviving copy of Tiglath-Pileser's royal annal, from the fifth year of his reign, takes the form of an octagonal prism recovered from the temple of Anu-Adad at Assur. In it, he presents himself as pious, strong and tireless in the defence of his people, the protection of his land and the expansion of his territory:

> I gained control over lands, mountains, towns, and princes who were hostile to Assur and I subdued their districts. I vied with 60 crowned heads and achieved victory over them in battle. I have neither rival in strife nor equal in conflict.

His was an eventful reign, with campaigns against the Arameans in the west as far as the Mediterranean, and into the mountains to the north aiming to secure trade routes. In the south he raided Babylonia, but his battle against Nebuchadnezzar was unsuccessful.

Numbers pervaded the text: there were nearly fifty different counts in Tiglath-Pileser's annal, including the kings he had fought with (sixty), the lands he had conquered (forty-two), the times he had crossed the Euphrates (twenty-eight), the numbers of his soldiers and his prisoners. The numbers ranged in size from a few dozen up to tens of thousands: 20,000 men-at-arms fought with, 6,000 hordes captured, 30 chariots taken, 300 families received, 2,000 captives taken, 12,000 troops conquered . . . Tributes imposed on the conquered were quantified, as was the immense plunder removed from conquered cities: mules, donkeys, oxen, rams, gold, silver, white bronze, tin, carnelian, lapis lazuli, agate, jewels . . . Tiglath-Pileser was particularly interested in counting the items brought back to Assur as gifts for the temples there: twenty-five 'gods' (presumably cult statues), copper kettles in various numbers, and on one occasion five bronze bathtubs. In similar vein, the inscription included a list of the king's hunting exploits, with a detailed enumeration of the wild bulls, elephants and lions killed or captured (plus, on the Mediterranean campaign, something called a 'nahiru fish' or 'sea horse': possibly a narwhal). The king thus spent his energies waging war on more than one front against the forces of chaos and darkness, on behalf of his people and at the command of the Assyrian gods.

The profusion of numbers in the royal inscriptions begs the question of how all the artefacts in question were counted – particularly for the larger numbers – and by whom, as well as who chose which items to count and which to describe with vaguer quantifiers like 'many', 'all' or 'as many as there were'. When the king's inscription did enumerate soldiers or plunder, the implication was presumably not that the king himself literally counted them. Who did?

Fortunately, imagery from later reigns can shed some light on the question. Several Assyrian royal reliefs show scribes involved in the counting process. One depicts an official dictating to a pair of scribes, on the roof of a building; one scribe has a tablet and stylus, the other a parchment or perhaps a papyrus. The same relief shows two soldiers chopping up a statue and others weighing seized goods: the whole scene gives an impression of some attention being paid to the quantification of what was captured. The scene with a pair of scribes is common in other such depictions:

Two scribes counting booty, in an Assyrian relief.

Typically a clean-shaven scribe holds a papyrus or parchment, next to a bearded one with a hinged wooden writing board, which opened out to reveal a waxed writing surface that could easily be erased for re-use (though other combinations of scribes are also known). The scribes are more often depicted pointing their styluses at the plunder or bodies they are counting than actually using them to write with.

In other words, the scribes were depicted in the very act of counting, using a stylus to help keep track of the sequence of objects while counting (in words? mentally?), ready to record the outcome in the form of number symbols. It is likely, then, that every important military action was followed by officials whose role included counting the plunder and collecting information about it.

These images of scribes on the battlefield give the impression – and were certainly intended to give the impression – that the items enumerated in royal inscriptions were really counted; that the numbers were neither mere guesses nor inventions. Most of the numbers are indeed plausible in their general size, though the larger ones were certainly rounded. Indeed, the Akkadian system of number symbols made it rather easy to round a number by simply leaving out the symbols for the lower denominations; most numbers in royal inscriptions in fact have only their first one or two digits specified. But they were not systematically distorted, as far as anyone can tell. The counting was real.

The monuments to Tiglath-Pileser I stood at the beginning of a long-lived tradition of Assyrian royal inscriptions, which continued down to the seventh century BCE, through periods of both eclipse and triumph for the aspirations of the Assyrian kings. Numbers always remained an important part of the inscriptions, and their uses expanded to include the duration of events, the size of cities built or conquered, their numbers of columns and the heights of their walls. In 722 BCE the Assyrian inscriptions reported the deportation of 27,290 people from a western neighbour, the kingdom of Israel.

The Assyrian royal inscriptions were placed in a range of locations. They were carved into stone stelae – free standing monuments – soon after the end of a military campaign and erected on its very spot. They were written on tablets, prisms or cylinders of clay and buried in the foundations of building projects. Their audience was in one sense the Assyrian court, the same circle of officers and scribes who produced the texts. In another sense it was the gods, or posterity, or the future kings who might dig up the foundation inscriptions, bringing the great deeds of their predecessors back into the light to check and even modify the record. For their purpose as propaganda, the use of numbers gives a sense of solidity to the information presented which mere words would not have done. They amplified the fact that, say, Tiglath-Pileser subdued 'a lot' of cities or brought home 'several' of their cult statues, imparting some precision – though admittedly without any possibility of verification – to what the text celebrated.

More than that, the royal inscriptions constituted both a new use of text and a new use of counting, and they illustrate the power of number words and number symbols to shift the way the world is described, perceived and categorised. In Sumer, number symbols made ten sheep into a single item, sixty containers of grain into a single possession, or the transfer of 3,600 vessels of oil into a single transaction. Tiglath-Pileser and his scribes and readers created a world in which conquering ten cities, subduing sixty rulers or carrying away five bronze bathtubs was – for a king – a single act.

Teianti: Counting coins

Egyptian Thebes; 28 February, 274 BCE. Ink scratching onto papyrus.

Teianti, daughter of Djeho and Tamin, bought a house (actually two houses) five years ago. She has just finished paying off the tax on the purchase. The scribe Esioh son of Djechensertais writes the receipt for her:

'There are 6 *kite*, making 3 *staters*, making 6 *kite* again which Teianti daughter of Djeho has paid as the 10% of the price of the house of Pabuche son of we-R' and of the house of Teihor daughter of Harsiesi, making 2 houses in all, which [they] sold being the 10% for the scribes of the bailiffs of Thebes. Written by the scribe of the land of Thebes Esioh son of Djechensertais, the scribe of the *phylae*, in year 9 *Typi* (day 1) of Pharaoh Ptolemy son of Ptolemy.'

Heir to an immense history, Egypt was by the 270s BCE ruled by a Greek-speaking dynasty whose founder, Ptolemy Soter, had been a general under Alexander the Great. There had been a Greek presence in Egypt since the seventh century BCE; the country was ruled by the Persians for periods in the fifth and fourth centuries. With its thousands of square kilometres of fertile land beside the Nile, Egypt was always a tempting prize to its neighbours. Its own urban tradition stretched back into the fourth millennium, with thousands of smaller towns and villages in addition to the great cities of Memphis and Thebes.

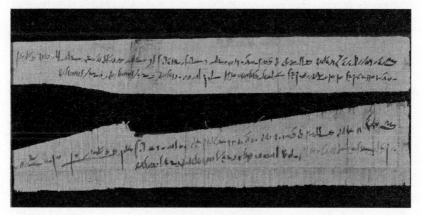

Teianti's tax receipt

Ptolemy I set up a literate administration run by scribes, centred on the new Greek cities. Many thousands of surviving documents testify to the details of local administration such as taxation and law enforcement. They also contain the annual land surveys that followed the Nile's yearly flood, reporting on the crops in the fields and enabling the crown, at least in principle, to estimate its income for the year.

Greek settlement was mainly in the Fayum, the former marsh near the Nile where drainage and irrigation made it possible to reclaim land and conjure into being a whole new province. By the later third century BCE a sixth of its population was Greek, including both military and civilians; there were villages where Greeks were the majority.

Yet in Thebes, 800 kilometres up the Nile, society remained overwhelmingly Egyptian: its temples were ancient and wealthy, as were the families of priests that served them. Here, the Ptolemies did little building work. Away from the main streets you would have seen mud-brick homes and shops lining narrow streets; most had courtyards and wells, with space to store animal feed. Such buildings functioned both as family homes and as sources of wealth, to use as security against loans or as dowries. Sometimes, collections of papers have survived, vividly recording networks of family and financial ties, of legal and social obligations.

The house that Teianti would eventually buy stood to the east of the river, in the city's northern district and in the vicinity of the Great Shrine of Amen. Its ownership is documented in a cache of papyri which trace it for more than fifty years, starting in the year 324 BCE: just two years before the Greeks took Egypt from the Persians. The first known owner, Djufachi, was a carpenter of the temple of Amen. In the ninth year of Alexander the Great, he apportioned his fairly substantial property – a set of buildings around a courtyard – among his children, leaving a section of the north wing beside the road to the younger two, Petechons and Phib. For a period, the two brothers lived in a single household, but they subsequently divided their property into two separate dwellings, part of which passed, through Petechons' marriage agreement, to his widow Taesi. Taesi used the house as security for a loan and subsequently sold it to the holder of the loan, an unrelated man named Pleehe who held the title 'Lector of the Ape'. It passed from him to his wife Teihor, who in turn sold it to Teianti.

Teianti's tax receipts survive for this and another property bought around the same time – 10 per cent was levied on such transfers of property – as well as a document showing that she then leased it to her sister. The papyrus trail ends in the year 274 BCE with her payment of the final instalment of the tax, a relatively careful and elaborate document in which the scribe took the trouble to give names in full, to spell out the nature of the tax being paid and to give the sum involved in both Egyptian *kite* and Greek *staters*. Its impression of business and bustle is surely not deceptive about this property; it had now been home to several different households within the world of Theban temple officials, and had changed hands five times in as many decades.

The language of this and similar documents from the same period is Egyptian, in a late form directly descended from the Egyptian of the previous 3,000 years. It is an Afro-Asiatic language, a member – at some distance – of the same group as the Semitic languages like Akkadian. The Egyptian language contained number words in a purely decimal system, with terms for the numbers from 1 to 9 and for the powers of 10 all the way up to 1 million. Those elements were used in combination (subject to some further complications) to specify numbers up to that limit.

In writing, the number words were rarely spelled out as words: in the hieroglyphic script, numbers were expressed using symbols, which also used a decimal system. There were seven signs, for 1, 10, 100 . . . 100,000, 1,000,000; each was written up to nine times in order to denote a number. The signs were placed in descending order of size. This was thus a very similar system to the Sumerian and Akkadian one, but without the complication of 60 as a base in addition to 10.

Egypt had a long tradition of calculation, of complex systems of weights and measures, and of scribes whose role and powers were very broadly comparable to those of Sumerian and Akkadian scribes, but were in many ways different. The scribes provided numerate expertise in contexts such as accounting, surveying, building and planning, and they are depicted in tomb decorations as well as physical models, in scenes such as supervising baking or brewing or inspecting goods of various kinds. The evidence is that over time some of the king's power came to be transferred to the scribes who worked in his name.

The hieroglyphic script was invented late in the fourth millennium BCE, around the same time the Sumerians first wrote their script. In fact it is possible – though not certain – that Egyptians received the general idea of writing words down from Sumerian models or reports of them, though their symbols were all their own. Mainly written on stone, the hieroglyphs were originally a set of about a thousand signs for things, plus a further set of twenty-four signs for specific consonant sounds; the two were used in combination. They were later simplified and made cursive for use on papyrus in the script called hieratic. And over time the hieratic writing of the Nile delta region became sufficiently distinct to be called a new script, namely demotic. Even more cursive and distant from the original hieroglyphs, demotic was used for over a thousand years, from the seventh century onwards.

In these later scripts, the number symbols not only looked different from those of hieroglyphic but worked to a different system. In the hieratic script, groups of number symbols came to be combined into single symbols, so that instead of writing '1' four times, say, you wrote '4' just once. Thus, for each number from 1 to 9, there was now a distinct sign. The multiples of ten from 10 to 90 also had their own set of distinct signs. In demotic at least, the multiples of 100 had another set of signs, as did the multiples of 1,000. The demotic number symbols were widely used in commerce, administration and legal documents, throughout the period when the Ptolemies ruled Egypt. It was in demotic that Teianti's tax receipt was written down; and the distinctive demotic number symbols are clearly visible there and in the other documents concerning her house and its previous owners.

Compared with the number systems that had preceded it in the region, this was something rather different: simpler in some respects but more complicated in others. Any number up to 9,999 could be written using no more than four signs, compared with the dozens of signs that some numbers needed in the hieroglyphic system. It was more compact, therefore, and faster to write. And a glance at a written number would tell you its approximate size in a way that a glance at a hieroglyphic number did not: a number involving four symbols was necessarily in the thousands, for instance.

But any attempt to see it as a straightforward improvement on its predecessors is liable to founder. The number of different signs users had to memorise was now relatively large. With nine different signs for the units, nine for the tens, nine for the hundreds and nine for the thousands, there were thirty-six different number symbols in use: more than the whole set of phonetic signs used to notate the consonants of the Egyptian language. (And this in a culture where a mistake could result in a beating: 'woe to your limbs' if you get things wrong, as one text about scribal life has it.) Furthermore, for the purpose of calculation, this was not a system whose symbols could be moved and accumulated intuitively like counters: it required you to memorise addition and multiplication tables, or to convert numbers into another representation – such as an abacus – in order to carry out even the simplest additions and subtractions.

Also, the system was not extensible in the way that a positional system – which reuses the same symbols for numbers of different sizes – can be. It had a fixed upper limit; there were no symbols for numbers above 9,000, and the largest number that could be represented was 9,999. If the number 10,000, or anything larger, needed to be written down, the rather cumbersome expedient was to write combinations of two lower signs, understood as being multiplied together. Like all ways of counting, in other words, the Egyptian demotic symbols were good for some tasks, but less good for others.

––––––

Teianti's collection of documents ended up in a pot in a tomb of the nineteenth dynasty which had been reused as a private dwelling. How they got there is obscure: deeds were essential to showing one's ownership of a property, meaning that collections of papyri like this would normally have been preserved with some care. Equally obscure is their excavation beginning in the nineteenth century CE, which has left the separate documents scattered among museums in the USA, England, France and Egypt. Teianti's final receipt for tax paid is now in London, at the British Museum.

The Mesopotamian and Egyptian ways of counting with symbols had been some of the longest-lived the world has ever seen: a continuous tradition of 3,000 years and more, with change and modification, but recognisably interrelated and springing from the same original impulse to turn simple tally marks into something with more structure and hence more power. They lasted because the civilisations they served lasted; because they worked; and because the region had little contact with other parts of the world – the Indus valley, China – that might have shown them alternatives.

Nevertheless, the archaeology of the region bears witness to both their rise and their fall. Both the scripts and their number symbols vanished from use during the first millennium CE. Cuneiform and Egyptian hieroglyphics became dead scripts, which no living person could read. In Egypt, the demotic script was replaced by a Greek-derived alphabet known as Coptic, and the demotic number symbols were abandoned in favour of Greek symbols imported – like the Ptolemies and their settler subjects – from the other side of the Mediterranean.

Yet the story of these systems of number symbols does not end here. Although the symbols were replaced, the structure of the demotic system persisted and passed into the Greek world, where it would travel widely and enjoy a long life even down to the present. There, though, it would have to compete with other systems of number symbols: and indeed with entirely different ways of counting and imagining numbers.

4

Counter culture from Athens
to the Atlantic

From the crossroads that is the Fertile Crescent, follow the coast of the Mediterranean: north and west. Through Anatolia, and on into Europe.

A very different branch of the story of counting sprang up in this part of the world. Everyone knows that the Greeks loved mathematics, that their philosophy privileged numbers and geometry and that their language of mathematical theorem and proof is still in use today. But how did they count? And, for that matter, what did they count? And what – if anything – of their counting practices did they bequeath to the later European traditions of the Roman Republic, the Roman Empire and the Latin Middle Ages?

There were number words and number symbols in Greece and throughout Europe from the classical period onwards; some were apparently based on Egyptian models. But the way of counting most closely associated with Greece, Rome, their empires and their successors is one that recalls the very first evidence for wild number lines: counters. The techniques of counters and counting boards dominated European experiences of counting and calculation from the fifth century BCE for more than two millennia, fading out as late as the seventeenth or eighteenth century. Efficient, practical and versatile, they would eventually fall victim to a new emphasis on writing and printing; and therefore on number symbols. But their story is a fascinating one, and it starts in Greece.

Philokleon: Counting votes

Athens, 325 BCE. A middle-aged man – call him Philokleon – rises early and makes his way to the law courts, at the northeast corner of the agora. Others have risen earlier still, and the ten entrances are already staffed and busy, would-be jurymen buzzing about like wasps. Up to ten courts sit on any one day, and each typically has either 201, 401 or 501 jurors, so there is quite a crowd. Philokleon uses the entrance marked with the name of his tribe, where he finds ten boxes lettered from alpha to kappa. He brings out a little boxwood plaque that identifies him as a member of this year's jury pool; it is stamped with a letter as well as his name and tribe, and he places it in the corresponding box: say box beta.

Once all the potential jurors from this tribe have come in and given up their plaques, the presiding archon signals slaves to pick up the boxes and shake them. The archon then picks out from each box one plaque, whose owner will both serve as a juror this day and perform some administrative functions.

The administrators (Philokleon is not picked) place all the plaques in a machine called a *kleroterion*: a tall marble slab with many slots, arranged in five columns; the plaques go into the slots in random order. Once they are in, the archon signals for a die – in fact probably a ball – to be released from the mechanism at the side, rather like some modern lottery machines. If the ball is white, the plaques in the top row of the kleroterion are picked for jury service this day; if it is black, they are not. A second ball is released to determine whether the plaques in the second row will be picked, and so on. Philokleon's plaque is in one of the rows picked today, so now he knows he will serve as a juror. He gets his plaque back. Next he has to find out in which court he will sit.

For this purpose, the court staff have already assigned a letter label to each of the courts, from lambda on. Philokleon and the other jurors go to a basket full of (probably fake) acorns and pick one out; each acorn has a letter on it. Say Philokleon gets mu, so he's assigned to court mu today. He gives up his plaque again to the archon, who puts it in a chest marked mu; he shows his acorn to another official, who gives him a staff whose colour matches that of the court to which he is assigned: say the blue one. Philokleon heads into that court.

Meanwhile, the magistrates are using the kleroterion again to decide who will preside over which court. Balls marked with the courts' colours are placed in the machine and are matched with balls naming the magistrates, who head to the courts as well.

Once everyone is assembled in the blue court, the game of tokens, counters and random selection is far from over. Philokleon gives up his acorn and staff to another official, who gives him a token marked with a letter of the alphabet, telling him in which section of the court to sit. Next, the box of plaques is used once more, to choose ten jurors to perform further administrative functions: to control the water clock, count the votes, and supervise payment at the end of the day. Only now can the spectators enter and the trial itself begin.

Once the pleas and evidence have been heard, a herald announces that the vote will be taken. There is no designated time for the jurors to confer, but in practice Philokleon can chat with his neighbours as they wait for yet another exchange of counters. Each juror gives up his seating token and claims in return two *psephoi*: literally pebbles, but in fact little brass wheels with an axle protruding on each side. Each walks to the voting urns, placing his 'guilty' ballot in one and his 'innocent' ballot in the other. The shape of the ballots, and possibly some sort of cover for the urns, makes it possible to be quite discreet about this if you want to. Philokleon puts his 'guilty' ballot in the bronze, 'valid' urn, the other in the wooden 'discard' urn. He receives in exchange a pay token.

The ballots are now tipped out of the bronze urn and counted by inserting them into a board with five hundred holes in it, putting the 'guilty' votes at one end and the 'innocent' at the other. The herald announces the result. If there is a question as to what penalty or damages to apply, a second vote has to be taken: the jurors receive their staffs or seating tokens again, hear the arguments, give up the seating tokens in exchange for ballots, vote again . . . Once all of these procedures are over, each juror finally receives a pay token.

The – perhaps long – day ends with a final exchange of tokens, as the jurors are paid. The box of plaques is brought in, they are picked out at random and each juror in turn gives up his pay token in exchange for his plaque and three *obols* (coins), concluding the day's flow of tokens. (Or not quite. One Athenian playwright has a joke in which the court runs out of small change and gives two jurors one larger coin between them, which they take to the fish market to exchange for smaller ones.)

Athenian ballot tokens.

The Greeks had a set of number words, inherited from their proto-Indo-European ancestors. The Proto-Indo-European *hoinos, duoh, treies, kwetuor* of perhaps 5,000 years before became the classical Greek *hen, duo, tria, tettara* in a development that is – as these things go – relatively easy to reconstruct. It was a purely decimal system: there were words for the numbers from 1 to 10, and larger numbers were built up from those by a combination of multiplication and addition: thus 43 would break down as 'four-ten-three' just as it does in English, although the separate elements were in some contexts so distorted as to be hard to recognise. Greek had some irregularities and alternatives in the -teens: for 18 and 19, particularly, it was possible to say what amounted to 'two from twenty' and 'one from twenty'. There were words for 1,000 (*chilios*) and 10,000 (*myrios*, a word originally meaning innumerable). Beyond that, there were no standard expressions.

Greek was also relatively well supplied with alternative series of number words to answer questions like 'which in a sequence' (first, second or third), 'how many times' (once, twice or thrice), 'into how many parts' (there's no real equivalent in English), or to express ideas like single, double and triple or twofold, threefold and fourfold. They used their spoken numbers throughout their culture: most enduringly in the work of Greek historians, poets and orators, who variously enumerated casualties, gifts, costs, debts, embezzlements and much more. Philokleon could have heard, a little earlier in the century, the orator Demosthenes' celebrated attack on his guardians and their handling of his father's estate:

> Gentlemen of the jury, my father left two factories, each of them a decent-sized business: thirty-two or -three sword makers worth five or six hundred drachmas apiece, the least of them worth not less than three hundred, from whom he got an income of three thousand drachmas per year free and clear; then sofa makers, twenty in number, who were security for a loan of four thousand drachmas, who brought

him twelve hundred drachmas free and clear and about a talent of silver lent out at twelve per cent, from which the interest every year came out to more than seven hundred drachmas. . . .

There is some question as to how much of this kind of thing even the most alert jury could follow. In fact, the uses of spoken numbers in Greek rhetoric ranged from pitiless forensic clarity with clear, correct and repeatable calculations, to much vaguer invocations of large numbers for the sake of their emotional appeal.

The Greeks also had a set – three sets, in fact – of number symbols. It was possible just to use the Greek letters, in the alphabet order inherited from the Phoenicians: *alpha, beta, gamma* . . . The books of the *Iliad* and the *Odyssey*, and of other classic Greek works, are to this day numbered using this system.

There was, second, a system established at least by the fifth century BCE which used letters as abbreviations for certain number words: M for *myrioi* (10,000), X for *chilioi* (1,0000, H for *hekaton* (100), Δ for *deka* (10). A simple vertical stroke like a tally mark was used for the ones. Each symbol was repeated as many times as needed. Long strings of the same symbol were avoided by using ∏ to stand for *pente* (5), wrapping it around an M, X, H or Δ as needed. (There is a widespread, though by no means universal, tendency for systems of number symbols to avoid ever needing more than four repetitions of the same symbol: ultimately perhaps a response to the limit of subitising, the fact that groups of more than four cannot be recognised at a glance but have to be explicitly counted.)

This system had the same structure as the Roman numerals, though the family relationship between the two systems is not completely clear: a common Etruscan ancestor may have been involved. In the Greek world it was widely used, particularly for notating sums of money. But from about the third century BCE it was edged out by a more concise alternative. This used all twenty-four of the Greek letters, plus three extra symbols, to signify the numbers from 1 to 9, the multiples of ten from 10 to 90 and the multiples of one hundred from 100 to 900. To extend it into the thousands, the symbols from 1 to 9 were reused with a special mark before them. Structurally identical to the Egyptian demotic system, it was quite possibly a translation of that system into Greek letters; it could have been encountered by Greek traders in the cities of the Nile valley. Despite needing a little more effort to learn than the other Greek systems, it became the dominant way of writing numbers in Greek. It had the advantages of nearly always using fewer symbols than the other system, of showing at a glance roughly how large the number was – how many digits it had – and being relatively easy to read out.

This alphabetic system of number symbols was used on public inscriptions throughout the Greek world, on coins and for purposes such as sections in texts or – later – pages in books. It survived until the fall of the Byzantine empire, and it could still be seen in the page numbers and section numbers of early printed Greek books in sixteenth-century Europe. Like the Roman numerals, it is still occasionally used to number pages today.

In ancient Greece, its most visible use was in public inscriptions, ranging from tribute assessments, records of loans, accounts of military expeditions, inventories and other accounts of institutions, to accounts of the auctions of confiscated property. This public, numerate documentation was a particularly Athenian habit, associated with Athenian democracy and its emphasis on – its display of – public accountability. Indeed, it came to an end with the tyrannies of the late fourth century BCE that ended Athenian democracy. Even if few perused the inventories and the accounts in detail, any citizen could do so in principle; some were even set out in column format as though to make them easier to read.

The accounts carved on stone and erected in public places were of course copies of accounts that also existed elsewhere, sometimes in fuller, more detailed versions, written in charcoal on whitened boards and stored in private archives. And they were the outcome of a process, more private still, of counting and arithmetic for which neither the ancient Athenian public nor modern historians have much direct evidence. Indeed, it does not seem that the Greeks used their number symbols for the dynamic processes of counting and arithmetic, to which purposes those symbols were very poorly suited. For that, they had another way.

————

Ancient Greece has well and wittily been called a counter culture, a culture in which most people's primary experience of and ideas about number – their ways of counting – involved not words or written symbols but small, graspable, manipulable objects: counters. (A world in which you could say 'number' and mean 'collection' or 'group', in turns of phrase that survive to this day: 'I want to be of that number'.) As Philokleon's day at the law courts illustrates, counters were everywhere in ancient Greece. Made of wood, clay, lead or bronze, or even of silver, they were used for games and for divination, they were used as tickets to political institutions or the theatre, and they may have been used in the distribution of grain.

And they were used for the workings of democracy. A system whose basic idea is majority decision-making must lean rather heavily on its practices of counting; must define them clearly and establish procedures to ensure they are transparent, reliable and trustworthy. There are rumours of a 'shouting vote' in neighbouring Sparta, while in the Athenian assembly a mere show of hands was used for most decisions. But in the law courts, decisions were more strictly quantitative, and the way Athenians selected, sorted and assigned individuals, as well as finally counted them, was to match them one-to-one with counters. Out of Athens' no more than perhaps 20,000 eligible citizens (those who were male and over thirty), a normal court day saw 2,000 of them engaged in the elaborate exchange of lots, tokens, counters and ballots. There were around two hundred court days in the year. The most important counter, the *psephos* or pebble, came to stand for the whole procedure, giving its name to the decision-making process and its outcomes, and even to the law courts themselves. One of Aristophanes' jokes about the subject has an obsessive devotee of the law courts waking up with his fingers stuck in the counter-holding position. The word was further transferred to the simple meaning of 'number', and it survives to this day in the term 'psephologist' for someone who studies voting patterns. Some Athenians took the plaques identifying them as jurors literally to their graves, arranging to be buried with them just as an earlier generation of Greeks had been buried with their swords and armour.

The complicated procedures of the late fourth century had of course developed from earlier, simpler ones. Originally, decisions were made by placing actual, literal pebbles in one of two open piles, by placing shells in urns, or even by making marks on wax tablets. The kleroterion was the culmination of procedures that had originally amounted simply to pulling a marked object (a pebble, a potsherd or a black or white bean) out of a vessel. The direction of change was towards greater secrecy for the individual, and greater transparency – less possibility for collusion and cheating – in the various random choices that had to be made.

The court's procedures, as they evolved, were also used for other things. The kleroterion was employed for other choices by lot, such as the annual selection of magistrates and the various other political officers: over a thousand individuals needed to be selected in total, and the machine could make the procedure much easier and more reliable. Surviving kleroteria are of various sizes with various numbers of vertical and horizontal rows, tailored to particular selection processes. There was even a process called *synklerosis* in which two of the machines were used together, to match random candidates to random offices.

Similarly, the use of counters to vote was adopted in the Athenian Council (of five hundred members) for cases where exact numbers were important, with a procedure probably similar to that in the courts. A bizarre variation occurred when the Council voted on the expulsion of a councillor, for which it used not pebbles nor bronze *psephoi* but olive leaves as counters. Meanwhile, if and when the full assembly of Athenian citizens chose to vote on the banishment of a citizen ('ostracism'), it used potsherds (*ostraka*) as counters.

The Greek world, of course, contained yet another kind of counter, one that was exchanged, handled and counted perhaps even more pervasively than the tokens and ballots of the political and legal instutions; that was counted, frequently, in both court speeches and public inscriptions. Coins.

The first coins in the world were struck in the early sixth century BCE in the vicinity of Ephesus, on the border between the Greek world and its neighbours to the east. They were an invention whose time had evidently come, and the making and use of coins spread through the Greek world and later the Greek empire, replacing (though not entirely) other ways of storing and exchanging value in the form of goods: cattle, cauldrons, corn, ingots and perhaps metal roasting spits. (The Greek word *obol*, the coin in which Philokleon was paid, embodied a memory of a time when goods were money, since its name literally meant a spit.)

Scraps of silver stamped with religious and civic emblems – turtles, owls, dolphins – and sometimes, later, with number symbols giving their date: these provided a new game of counters for Greeks to play, a new sphere in which to participate and a new way to measure your worth. They made wealth just as handy, tangible and countable as votes.

The Greek states, their institutions, and the concept of an all-purpose money developed hand in hand: nowhere more so than in Athens, whose currency and weight standards dominated the Greek world through the fifth and fourth centuries, and where talking and writing about finance – pay, trade, gifts, bribery, theft, tribute – became ubiquitous, creating new uses for the Greek number words and number symbols. When individual wealth had become a source of both complaint and excuse, it was not unusual for those pleading in the courts to go to some lengths to complain of their opponents' wealth, to deny being wealthy themselves, evidently in the hope of playing on jurors' presumed prejudices. When everything had its price, all a politician had to do to win over the public was lower the price of sardines (so it was said). When coins were everywhere, there was a need for whole new professions to handle them: bankers, money changers and coin testers.

When Philokleon received his three obols at the end of a long day of exchanging one counter for another, he could promptly go to the market and exchange them for goods. Items that had once been bartered against one another could now be conveniently counted on a common scale using silver counters. And not just everyday items like food or clothing. By embodying a universal standard of value, coins created equivalences between different spheres where none existed before: marriage contracts, court fines, gifts and pay, for instance, were all now in a sense interchangeable; they could be counted off against one another and placed into mutual correspondence. A fragment from the mid-fourth century reports that in the agora you could now buy 'figs, marshals of the court, grape bunches, turnips, pears, apples, witnesses, roses, medlars, haggis, honeycombs, chickpeas, lawsuits, beestings, curds, myrtle, allotment machines, blue cloth, lambs, water clocks, laws, indictments . . .' The strange alchemy of money seemed capable of transmuting anything into anything else.

Marcus Aurelius: Counting years

Rome, 176 CE. Emperor Marcus Aurelius has been away from the city for years, on the Marcomannic campaigns beyond the Danube. He addresses the Roman people on his return. They are entitled to receive a gold piece, an *aureus*, for each year he has been absent:

'While he was saying, among other things, that he had been absent many years, they cried out, 'eight,' and indicated this also with their hands, in order that they might receive that number of gold pieces for a banquet. He smiled and also said 'eight'; and later he distributed to them eight hundred *sesterces* apiece, a larger amount than they had ever received before.'

While systems of counting words developed to more sophisticated internal structures and with the capacity to represent ever larger numbers; while the use of counters burgeoned into the complexities of the Greek voting systems as well as turning into coins; and while number symbols attained something of their modern range and flexibility across the Near East and the Mediterranean world, what happened to counting on the fingers? This way of counting, too, is capable of development, change and sophistication. There have been many systems of counting gestures more complex than simply extending or folding one finger at a time to count up to five. One of the best-documented and longest-lived was that of ancient Rome.

Its origins are lost: the earliest firm evidence is from Rome in the third century BCE, and it is quite unclear whether it existed earlier in Greece or elsewhere. Though Greek has an expression for counting on the fingers, and Greek literature has the odd mention of sums so easy you can do them on your fingers, specific reference to the elaborate Roman system is lacking. An Egyptian origin has been suggested; a Persian connection is hinted at in some sources, but evidence is lacking.

How to count in gestures, from a manuscript of Bede's book

The system provided hand signs enabling counting up to 99 on one hand, up to 9,999 using both hands. Invented apparently in a region where number words and number symbols counted in tens, it was not surprisingly a decimal system. Numbers from 1 to 9 used the little finger, ring finger and middle finger of the left hand, bent down either fully or partly and in various combinations. The multiples of ten from 10 to 90 were shown with the thumb and index finger of the same hand. For larger numbers, the same set of signs was used on the right hand to show the hundreds and thousands: the right thumb and index finger took the hundreds, and the right middle, ring and little fingers the thousands.

Throughout the period of the classic Roman authors, the finger-counting system was well enough known to serve as the basis for jokes and double meanings. Quintilian, in the first century CE, mentioned that orators were expected to be fluent in the finger numbers so as not to give a poor impression of their training. Juvenal, a little later, was able to imply that someone was very old by saying that he counted his age on his right hand. Catullus in the previous century could say 'give me a thousand kisses, then a hundred, then another thousand, then a second hundred . . .' and be understood by at least some of his readers to be providing instructions that led to a middle-finger gesture on the right hand (yes, it was already rude in those days; the finger itself was known as the *impudicus*, the rude one). Up to two dozen Latin authors might be cited for similar hints or discussions of the gestures accompanying particular numbers, from Plautus in the third century BCE to Cassiodorus nearly a thousand years later.

The gestures were used to show a count to others, or to keep track of a count for yourself; mention was made of their use in the law courts, the arena, the public assembly and the church. Background noise and distance between people wanting to communicate would have made them useful in those situations and many others, and they persisted alongside spoken and written numbers because they were good for different things.

More vivid than textual references are the surviving depictions

of number gestures. Counting-house scenes or pictures of negoti-
ation about prices – with the participants showing each other
numbers in gesture – appear in carvings, reliefs and mosaics. What
seem to have been gaming pieces, with the hand gestures depicted
on one side of them, turn up in archaeological contexts: a full set
apparently showed the numbers from 1 to 15.

More symbolic uses also occurred. The legendary second king
of Rome dedicated a statue of Janus which displayed the number
365 on its two hands: the number of days in the year, apt for
the god of time and duration. A late Roman tradition placed
pictures of the hand signs on sarcophagi, showing the number
52: the number of weeks in the year and a symbol of the cycle
of life.

Several ancient sources refer not just to counting on the fingers
in this way but also to performing calculations with numbers in
the same form. But no surviving source explains in any detail how
it worked. It would have been possible in principle to memorise
the addition and multiplication tables in the form of hand gestures,
but it is perhaps more likely that the fingers were used primarily
for simpler computations, or to keep track of intermediate stages
in calculations being performed mentally.

The Roman finger-counting system was hugely long-lived. It
survived the end of the Western Roman Empire and was still in
widespread use well into the Middle Ages. In medieval contexts,
it provided a way to give gestures and postures an additional layer
of meaning: to insert numbers and number symbolism into texts
and pictures. A second-century Gnostic text found at Nag Hammadi
contains an exposition of the New Testament parable of the lost
sheep which makes use of the fact that the ninety-nine sheep –
who never go astray – are counted on the left hand, but the full

count of one hundred sheep – after the stray has been recovered – passes to the more auspicious right hand. Similar exegetical moves can be found in the sermons of several Christian writers including St Augustine, addressing congregations of no remarkable education who were nevertheless expected to see the point at once, to be familiar with the Roman finger-count.

In fact, like most things that 'everyone knows', the finger-counting system was seldom explained or described in writing: at least until much later, when it was in danger of dying out and had become more specialised knowledge. The earliest complete descriptions of the system – and therefore the basis for interpreting allusions and uses of it from the Roman period – date from the late seventh and early eighth centuries: a short pamphlet called the 'Roman computation' whose earliest copy probably dates to 688, and a better-known discussion by the Venerable Bede, monk of Jarrow in northern England, from a few decades later. Bede addressed the finger-counting system in the first chapter of his book on calendar calculations, *The Reckoning of Time*. Writing in Latin for his fellow monks, he praised the system and implicitly admitted that by this time it faced some resistance: it was 'very useful and easy', he said; one should not 'despise' it or 'treat it lightly'. Presumably some did.

While the gestures Bede described seem to be the same as those used by earlier Latin writers, with just a few minor alterations over the centuries, both he and the 'Roman computation' pamphlet described a way to extend the count beyond 9,999 that has no parallel in any earlier source, and that may have been a medieval innovation. For the ten thousands you would place the right hand at various spots on the body in turn: head, throat, breast, side, stomach, groin, thigh. The left hand was then similarly employed to count the hundred thousands, while one million had a special gesture of its own with the two hands held at the sides of the face. The details differ from one copy of Bede's text to another, and it

is unclear how much this extended system was ever used in practice.

Bede's book – and in fact separate copies of its first chapter –
circulated widely, and a large number of copies survive to this day.
Many were illustrated to show what the finger signs should look
like. Despite this, evidence of the system in actual use in the Latin
world becomes sparse after Bede's time. By the time late medieval
and Renaissance writers – such as the mathematicians Leonardo
of Pisa in the thirteenth century and Luca Pacioli in the fifteenth
– described the system in their books, it seems to have been of
merely antiquarian interest. The same is true of printed versions
of Bede's *Reckoning of Time*, which appeared from 1525 onwards.

Meanwhile, though, the system was also in use outside the Latin
world, in the other inheritors of parts of the Roman Empire. A
fourteenth-century Greek author at Smyrna described a version of
the system. His treatise also exists in an Arabic version, and Arabic
and Persian texts contain references to the system down to about
the seventeenth century.

During the same period, there is also visual evidence for a
continuing interest in the number signs in Europe. Frescoes and
altarpieces continued, occasionally, to incorporate figures making
the hand signs for numbers of significance to the scene in which
they stood, down to the twelfth century if not later. The ability of
viewers to interpret the gestures could eventually no longer be
relied on, and the practice dropped out of use.

Blanche of Castile: Counting with silver

To return to 'counter culture'. How did Greeks and Romans perform calculations, or handle more complex sets of numbers than simple counts?

The Greek and Roman systems of number symbols – like the Near Eastern systems that preceded them – were better suited to recording the outcome of a count than to performing it, and like most ways of writing down numbers they were particularly poorly suited to arithmetical calculation. Instead, for counting and calculation – as opposed to long-term record – people turned again to their counters, and to a key element of their 'counter culture': the counting board.

The counting board, taken broadly, is a technology that seems to have been independently invented more than once, and in some of its forms it has left evidence that is frustratingly scant or ambiguous. It is generally assumed that the scribes of the ancient Near East used a board of some kind for their calculations, but no such boards or even descriptions have survived. In Greece, though, the counting board was a central technique for computation, and it is both mentioned in literature and found in archaeological sites. Wherever numbers needed to be added, subtracted or otherwise reckoned with, from the courts to the banks to the practice of engineers, surveyors and architects, the counting board was used.

It consisted of very little. Any flat surface would do, together with a set of objects to use as counters. A set of lines would be drawn or engraved, optionally labelled, and the counters placed on them and moved as necessary. It has been well said that this was more a state of mind than an artefact. The name *abax* was used both for counting boards and for gaming tables and other items of furniture, including sideboards and even trenchers or plates. If it is correct, as often claimed, that it relates to a word in the Semitic languages meaning dust (*abq*), its first meaning may in fact have been a dust-covered writing board.

Around thirty surfaces marked for use as counting boards have survived from the ancient Greek world, though it is not always easy to be certain whether they are counting boards or game boards (or were used for both functions). They range from large marble tables to repurposed roof tiles; at the time, in fact, the majority of counting boards would surely have been wooden, and have therefore not survived. The counters could have been ordinary pebbles or even coins.

The lines, whether labelled or not, were used in a basically decimal system: there was a ones line, a tens line, a hundreds line, and so on. There were also, closely matching the older Greek way of writing numbers, a fives line, a fifties line, a five-hundreds line, and so on. Some boards interspersed these so that the values of the lines were 1 – 5 – 10 – 50 – 100 – 500; others simply used the top halves of the ones, tens and hundreds lines for fives, fifties and five hundreds. The effect, in either case, was that you never needed to deal with a large group of counters on any one part of the board: if the units line got to five counters you replaced them with one on the fives line; if the fives line got to two counters you replaced them with one on the tens line, and so on. There was a much-used ancient quip, that kings move courtiers around like pebbles on the counting board, making them now worth more, now less. Groups of more than four counters on one line were avoided, just as repetitions of one symbol more than four times were avoided in the Greek and Roman systems of number symbols: and perhaps for the same reason, that such groups lie beyond the human ability to count at a glance. Many Greek boards, finally, boasted a special set of lines for fractions, whose different values followed the denominations of the coinage system.

———

The counting board was just as ubiquitous in ancient Rome as it had been in Greece, and has left still more sparse evidence: a handful of references in literature, a few (five) archaeological finds and no actual description of how it was used. The word *calculus* literally means a pebble in Latin, but became transferred to mean a reckoning, computation or calculation; its descendants today provide words in several languages with similar meanings: *calculate* in English, *calcolare* in Italian, *calcular* in Spanish.

Physically, the Roman counting boards that have been found differ strikingly from the Greek examples. They are made of bronze, they are small – around eight by twelve centimetres – and their counters are permanently bound to them. That is, the 'counters' take the form of metal studs held in narrow slots in the metal surface of the board, bringing the whole device close to what is normally meant by the – vexingly ambiguous – English word 'abacus'. The arrangement of the slots is that of the more compact Greek boards, with the column for fives above that for ones, for fifties above that for tens, and so on. This structure of fives and tens of course closely matched that of the Roman numerals, making it easy to translate a number from board to writing or vice versa: exactly one written symbol was required for each counter in play. (Roman numerals of the ancient period generally represented 4 as IIII: the convention of writing it as IV is a later one.)

A third-century bas-relief showing two Romans moving counters on a table is the best evidence for the existence of the counting board in the later Roman Empire; it is generally assumed that its use continued widely until the fall of the Western empire in the late fifth century. Thereafter, though, evidence – both literary and archaeological – falls silent until a clutch of texts from the tenth and eleventh centuries show that the counting board was alive and well in the Latin Middle Ages. There was a vogue at that time for learned or at least quasi-learned expositions of the counting board, its operations and the uses to which it could be put in learning about number theory, geometry or even logic. Some involved modified forms of the board and counters, but the whole phenomenon probably amounted to an attempt to rescue for respectability and learned study a device which had never been out of the hands of ordinary people; or indeed the hands of their rulers.

Paris, in the 1240s. A learned clerk sits at his counting board and reckons, showing his queen how much has been raised in revenue, how much spent, what achieved and how much yet to be achieved in a lavish monastic building project. Her name is Blanche, and she is in her fifties. Keen-eyed, alert, interested, and fluent in French, Spanish and Latin, she quizzes him closely, following the reckoning carefully. Possibly she checks the account by doing her own calculations on the clerk's counting board or on one of her own. Specially struck counters bearing the royal arms fly under their fingers.

Blanche of Castile was born in 1188, and lived one of the more adventurous lives of the period. She was the granddaughter of Henry II of England and his queen Eleanor of Aquitaine; Richard I 'the Lionheart' was an uncle, as was King John of England. But on her father's side, she was a Castilian princess, the daughter of Alfonso VIII.

Blanche of Castile.

At the turn of the century, aged twelve, she was married to Prince Louis, heir to the French throne; she never returned to Castile. Louis acceded to the French throne in 1223 but died after only three years of rule. Blanche wept so hard and so long that her courtiers feared for her sanity. There followed three decades during which joint rule with her son, another Louis, alternated with periods when Blanche was for all practical purposes the sole ruler of France. For the eight remaining years of Louis' minority, during his illness in 1244–5 and after he left on crusade in 1248, her power was complete.

At all times, the dowager queen wielded diplomatic power, sat in judgement and organised some of the major events of court life. Her contemporary, the scribe and scholar Matthew Paris, reckoned her the greatest lady on Earth. A recent biographer calls her one of the most imaginative and successful rulers of medieval Europe, 'remarkably adept at ensuring the people did what she wanted', both at settling high-level diplomatic disputes and dealing with the everyday business of government:

> She reacted courageously to challenge and opposition, whether from the church, Paris masters or the barons. She used the full range of coercive powers available to the ruler. She raised armies; she sat in judgement; she issued at least one kingdom-wide ordinance . . . and she was a determined negotiator.

Beyond the court, the events of her life included the collapse of the English kings' control of much of modern France, magnate revolts in France as well as in Spain and England, and new episodes of conflict in the Holy Land: Saladin captured Jerusalem in the year of Blanche's birth. When she was in her forties, Europe began to hear news of the Mongol threat from Central Asia.

As well as acting on the largest possible stage, Blanche also had a reputation for control, discipline and excellence in more intimate matters. She was famed for her piety: her son Louis became one of the patron saints of France. In this, the period of cathedral building at Bourges and Chartres, she was a noted, even a prolific, patron of art and architecture, transforming the dull court she entered in 1200 into the cultural capital of Europe. Her dowry and her royal grants gave her land in various parts of France as well as huge sums in cash, and Blanche's personal income was stupendous even in a period conspicuous for the accumulation of wealth in the hands of a few: by the 1240s it was around 45,000 livres each year. (For comparison, her son would spend (only) 40,000 livres building the Saint-Chapelle in Paris.) Three Cistercian monasteries and one castle of her foundation survive in whole or in part. Psalters, Bibles, crosiers and stained glass from her patronage also bear witness to the magnificence of her taste and her gifts. Even her personal seal depicted her as a woman of 'fashion and charm'.

All of this – the public and the personal – took organisation, and the government over which she presided had a new administrative complexity compared with its predecessors. The royal household comprised something like three hundred staff: from men-at-arms to 'valets of the dogs'; from sommeliers and fruiterers to ladies in waiting; and including two dozen or more clerks, most of them educated in Paris. The clerks controlled expenditure, although during Blanche's lifetime the household was a complicated patchwork of different responsibilities, with various members of the household allowed to authorise spending.

The accounts of Blanche's court show the range of these payments over a highly complex set of purchases, payments and loans: payments to personnel, the expense of visitors, gifts, patronage and alms (Saint Louis himself worried that his mother's almsgiving was excessive), the organisation and funding of travel and of course the actual upkeep of houses and castles, plus recreation, food and clothing: everything from children's gloves, parchment and bookbinding to gifts to ambassadors, entertainment at royal weddings and payments sent out to the crusaders at Damietta.

Blanche must have spent much time discussing the details of food, clothing, jewellery and furnishings, the maintenance of buildings, hunting and music, as well as the logistics of court life, travel and entertainment. And she evidently spent time personally overseeing the household's accounts, authorising many expenses herself. For at least one of her monastic foundations she worked closely with one of her clerks to establish what income was available from the designated lands and what the work would cost.

It is perhaps no surprise, then, that Blanche was personally associated with a remarkable innovation in the technology of counter and board. From time immemorial, counting boards had been stocked with pebbles, potsherds or perhaps coins as counters. But early in the thirteenth century, there appear at the French court *jetons*: counters specially struck for use on counting boards. The word derives from *jeter*: to throw or perhaps to push, in other words to push counters around on the board.

The earliest known jeton bears the arms of Blanche of Castile: the fleur de lys on one side, the castle denoting Castile on the other. It was one of a set struck for use at her court: for use, that is, by her clerks or perhaps by the queen herself. Later descriptions show noble lords following along the reckonings of their officials on their own counting table and with their own counters. It was very much of a piece with the rich carpets, clothes and jewels, the lavish books and crosiers commissioned by Blanche, to have special counters for use in her hands or in her presence.

One of Blanche's jetons.

From this period and over the next few centuries, the evidence for counters and counting boards burgeons. There are references in literature, from Chaucer's 'counting-bord' to the Shakespearean insult 'counter-caster' in English alone. Counting tables are mentioned in wills and inventories, and from the sixteenth century onwards, there are several actual counting boards that have survived to the present day.

Some were elaborate and expensive pieces of furniture, sometimes with special drawers to hold the counters and capable of being folded up when not in use; some had surfaces with two or even three areas ready-marked for counting and calculation, so that several people could work side by side. Others were designed as ordinary tables that could easily be used for other purposes when calculation was not taking place. The most minimal form of the medieval counting 'board' was a simple cloth, marked with the required set of lines and labels and ready to be placed on any table of suitable size. The counting house of the English royal household used such a cloth: it was green, and the office came to be called the Court of Green Cloth. In several languages, table-tops in shops or at home are still called the counter, *comptoir* or similar, reflecting the fact that such surfaces once regularly functioned as counting boards. (Words like *to count* and *a counter* in English, *un comptoir* (a counter) in French and *il conto* (the bill) in Italian all derive from the Latin *computare*, meaning 'to calculate'.)

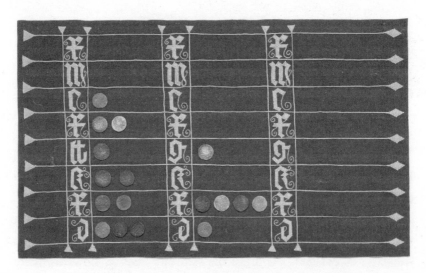

Counting cloth from Munich (replica).

In such forms, counting boards were in regular use by merchants and administrators of all kinds across the Latin world, from royal courts and cathedral schools to tradesmen, merchants, lawyers and clergy, as well as in private homes of various sizes and styles.

When the first printed descriptions of counting boards appeared in the sixteenth century, almost nothing seemed to have changed in the form and use of the device since the 'counter culture' of ancient Greece. One of the first was Jacob Köbel's 'Reckoning Book', written in German in 1514 and published in Augsburg. The title page sported a picture of a mistress and a maid at a counting table, presumably settling weekly or monthly accounts. The table was marked with lines and, in one place, a cross; coin-sized counters were scattered across it. Köbel depicted and described a device whose use would have been perfectly comprehensible 2,000 years before. The counting board still consisted essentially of a flat surface and the skills to put it to use.

Köbel began by describing the process of turning a surface into a counting board by drawing lines and, optionally, labelling them; he seems to have thought pre-marked boards were the exception rather than the rule. On Köbel's board the lines were horizontal to the user rather than vertical, a ninety-degree rotation compared with the Greek boards. But the sequence of lines denoting fives and tens was exactly the same as before. As with the Greek and Roman counting boards, beyond the ones line there was a set of further lines for fractions, reflecting the denominations of currency. Like some of his contemporaries, Köbel suggested the very slight innovation of putting a star next to the thousands line to help its quick recognition. He also suggested that, rather than labelling the lines using writing, you simply lay a coin of the appropriate value at the end of each one. One of the beauties of the counting board was that no essential modifications of its form or its procedures were needed in order to calculate with non-decimal systems of currency, weight or measure: you could simply assign, say, ounces to one line and pounds to the next, or inches, feet and yards, or any other system that was required.

Scene at the counting board, from Köbel's title page.

Köbel's procedures always started with the smallest denominations. For the ones in your number, set out counters on the ones line; for the tens, counters on the tens line, and so on. To add another number to this first one, start with the ones again and place extra counters corresponding to the ones in the new number; then for the tens and for the hundreds. If at any point you find you have placed five counters on a ones (or tens, or hundreds) line, you must remove them and substitute a single counter in the adjacent fives (or fifties, or five hundreds) line. If there are two counters on a fives (fifties, five hundreds) line, remove them and substitute a single counter on the tens (hundreds, thousands) line.

Subtraction was scarcely more difficult, although in order to have counters to take away from a given column you might briefly have to substitute five 'ones' counters for a single 'fives' counter: or the equivalent elsewhere on the board. To multiply or divide, though, it was necessary to have memorised the multiplication table, and to work with fractions you had to understand the system of coin denominations in operation.

Köbel also mentioned that it was possible to divide the board into more than one section by adding vertical lines, gaining the possibility of representing two or more numbers simultaneously, perhaps for the purpose of keeping track of a complicated calculation. Other writers advocated another slight change: using the spaces between the lines, rather than the lines themselves, to place the counters. Functionally this was a trivial shift, making no real difference to the function of the counting board; but it changed the physical look of (some) boards into a set of – possibly different-coloured – bands. If this was combined with the lengthways division of the whole board, the whole would start to look something like a chess board. The special form of board used for English public accounts was indeed known as the 'exchequer', a word from the same root as 'chess'.

Meanwhile, Blanche's innovation – counters specially struck for use on the counting board – showed a similar, and surprising, longevity. It could have been a mere fad, the toy of a particular queen with a particular interest in administration and accounts. But it was not. Jetons were struck for every subsequent French monarch, from the thirteenth century until the Revolution in 1789. From the royal court itself, the use of specially struck jetons spread to the various departments of the royal household and beyond: in due course the orders of chivalry, the assembly of French clergy, the households of the queen and the dauphin, and the treasury in its various departments all produced their own series of jetons. Provincial administrations and personages had their own jetons struck, and it seems that not only the royal mint in Paris but mints around the country – Tours, Dijon, Toulouse, Grenoble, Arras, Laon, Valenciennes – were also producing jetons at the peak of their popularity. They were made by the same simple process – hammer striking – as early coinage.

Early examples were in brass, but by the fourteenth century silver was in use, and subsequently gold, meaning that jetons had real value as bullion. The custom of giving a set of one hundred jetons to certain officials each year, nominally for use in their work, took on the character of a supplement to their salaries: something like an annual bonus. It may have been possible either to sell the jetons back to the treasury for their value as metal, or perhaps to commission their transformation – melting, casting and striking – directly into coin. One of the things that makes jetons intriguing is how closely they superficially resembled coins while always remaining distinct from them. Their shape and size and the choice of metals made them coinlike in all but their detailed design. And their designs tended towards the celebration of notables and their exploits, with a proliferation in later centuries of portraits, blazons and mottoes: just like coins. Some bore as part of their design a statement that they were not coin of the realm, but the practice of passing them off as money nevertheless led to the long-lived proverb 'false as a jeton'.

The peak of jetons' range and complexity in France came under Louis XIV in the seventeenth century; in the 1680s, the jetons struck for the central institutions of the French government numbered 1,200 in gold and a staggering 48,000 in silver. The issuing of jetons continued sporadically for another century, up to about 1780. But their close connection with the royal courts and institutions meant that they entirely ceased to be produced after the Revolution, and that the overwhelming majority of jetons extant at that point – with their royal and noble portraits and inscriptions – were swiftly melted down.

But for the centuries between Blanche and the last Louis, there seems no reason to doubt that actual computation in royal households, financial departments and elsewhere in the government was carried out using the jetons struck in such numbers for that purpose: that jetons really were preferred as counters, even though functionally they represented no improvement whatever over the use of pebbles or coins for the purpose.

Jetons also became popular outside France. There were important centres of their manufacture and use among the merchant and banking community of northern Italy – the Lombards who exported their goods and their practices to almost every corner of Europe – and subsequently at Nuremberg, whose 'reckoning pennies' came to be well known and widely exported throughout the sixteenth and seventeenth centuries. Some were self-referentially stamped with images of an accountant seated at a board using jetons.

From the seventeenth century onwards, jetons gained siblings in other kinds of counters with overlapping functions and physical characteristics. The habit of giving away sets of jetons as a new year bonus to employees or a birthday gift to friends spread to institutions such as religious establishments, university faculties and learned societies, which took to giving jeton-like tokens to their members in exchange for attendance at meetings or the performance of other duties. Lead or brass were often used (these tokens are sometimes distinguished from jetons by the term *meraux*). As quasi-money which could in some circumstances be – illegally or semi-legally – used as small change, or might be legitimately handed over to the issuing authority in return for money, they have some overlap with jetons proper. The difference was that there was never any suggestion that they might be used for calculation on a board.

Further afield still from the central purpose of jetons, there are references to their use for marking points at games, a safer alternative to putting actual money on the table during play. In due course, tokens were made for this specific purpose, sometimes in special shapes so that they could not easily be confused with other tokens or coins: the ancestors of modern-day poker chips. And the reluctance of certain governments – notably the English – to strike small-denomination coins led to several periods when merchants produced their own, in copper, brass or lead, counters that served the dual functions of small change and advertising. Tens of thousands of different types were produced, and they were manufactured on an industrial scale. There came to be a fad for collecting these 'merchants' tokens', leading some makers to issue commemorative or decorative tokens specifically designed for collection. Some collectors even had their own tokens specially manufactured to give to their friends.

Although there are certainly separate categories of object here, it also seems certain that between jetons and their siblings there was a long history of influence, imitation and – surely in some cases – crossover use. There is a sense, then, that by the seventeenth century the counter-saturated culture of classical Athens had been recreated in northern Europe, with a range and complexity of tokens in circulation as great as anything ever seen anywhere. And, consequently, that for most people, a primary experience of number and counting came through the fingers, from counters and their manipulation.

By this time, though, a new leaf was being turned as far as counting and calculating in Europe were concerned. Blanche's clerks used Roman numerals only to record the outcome of their calculations; but new techniques of calculation on paper had become ever more visible since the high Middle Ages, and it was during the sixteenth and seventeenth centuries that they at last took over from counters and counting boards. Today the number symbols are called the Arabic numerals: but their ultimate source was India.

5

Number symbols from India

Number symbols were a crucial addition to the human repertoire of ways of counting, when they were invented in Sumer and, concurrently or soon after, in Egypt. But a majority of the people alive today are familiar with one set of number symbols above all: the ones that look like this – *0 1 2 3 4 5 6 7 8 9* – and that obey a special set of rules about their positions and their values. They are particularly associated with calculation, but their spread is certainly one of the main branches on the story of human counting.

They were first used in texts in Sanskrit in the first millennium CE, but theirs has been a remarkable journey from language to language and from country to country. By the twelfth century they had a history of hundreds of years in South Asia and were well established in the Arabic-speaking world as well. And they were just beginning to appear in texts written in Latin, too. Over the following centuries they would conquer Europe and become a key element of modern accounting, of data-gathering and statistical analysis: techniques at the heart of modernity.

Was there anything inevitable about the rise of the Indian number symbols? Surely not: South Asia has other systems of number symbols with the same structure, that show no sign of spreading across the world. Happenstance, as in almost any historical process, played a leading part. Are they the last word in human counting? Again, surely not: they have not yet survived as long as the cuneiform or the hieroglyphic number symbols, and in the last few decades they have lost a great deal of ground to binary codes as a way of counting, of representing and manipulating numbers. There seems no reason to suppose that the process of growth and change in human counting is at an end. Their importance, and their interest, lies instead in the myriad places they have been and things they have done over the last thousand years and more. And their story begins in India.

Bhaskara II: Brahmi numerals

South Asia is a region where language, culture and counting have deep roots. Indo-European languages have been spoken in the Indian subcontinent from at least the second millennium BCE; the most prestigious was Sanskrit. Urbanisation in the first millennium and the exchange of goods and ideas with India's neighbours – Greece, Mesopotamia and China – contributed to a thriving and numerate culture.

The earliest surviving texts in an Indian language are the Vedas, in an archaic form of Sanskrit, dating from perhaps 1500–500 BCE; around 500 BCE, Sanskrit was codified and effectively frozen as a learned language. Its descendants include many languages and dialects spoken today, including Hindi, Gujarati and Bengali.

Evidence for counting in the region includes a tradition of wooden tallies which survived into the twentieth century, and references in ancient texts to a counting board on which clay counting pieces took different values depending on their position. Tallies and boards, like the Sanskrit language itself, adopted a decimal system. Inherited from Proto-Indo-European, like those of the European languages, the words were *eka, dva, tri, catur, panca* . . . (Similarities with number words in Greek and Latin were in fact one of the first pieces of evidence to be collected showing the relationship of those languages with Sanskrit.) Unlike many languages, Sanskrit has a wholly regular decimal system of number words, with a separate word for each new power of 10 (contrast English, where only *ten, hundred, thousand* and the higher powers of 1,000 have distinct words; other powers of 10 are named using compound terms like 'ten thousand').

Counting words occurred in some of the Vedic hymns ('You, radiant Agni, are the lord of all offerings; you are the distributor of thousands, hundreds, tens of good things'). In some cases long sequences of powers of ten were incorporated:

> Hail to a hundred,
> hail to a thousand,
> hail to *ayuta* [ten thousand],
> hail to *niyuta* [hundred thousand],
> hail to *prayuta* [million],
> hail to *arbuda* [ten million],
> hail to *nyarbuda* [hundred million],
> hail to *samudra* [billion],
> hail to *madhya* [ten billion],
> hail to *anta* [hundred billion],
> hail to *parardha* [trillion],
> hail to the dawn,
> hail to the daybreak,
> hail to the world,
> hail to all.

Although Sanskrit was primarily prized as a language of oral poetry, it gained a written form during the first millennium BCE. The climate of the region is not kind to perishable writing surfaces, but vernacular dialects were being inscribed on stone in the third century and there are inscriptions in Sanskrit itself by the first century BCE, with manuscripts surviving from around 150 CE onwards.

From the first half of the first millennium CE, there is also increasing evidence for the use of number and calculation in Sanskrit culture, in such applications as finding the number of possible verse metres, laying out the brick altars for fire sacrifices, or keeping track of the calendar for ritual purposes. It was astronomy (and astrology, and the study of calendars) that ultimately prompted the most mathematical work in Sanskrit. The discipline – one of the traditional 'limbs' of Vedic learning – flourished throughout the first millennium, producing a wealth of texts of increasing complexity and sophistication. The core of each text was in verse, but lengthy prose commentaries explained and amplified the rules of calculation and procedure.

One of the most striking features of Sanskrit astronomy is the model of space and time it used, which involved immense spans of space and cycles of time. Historian Kim Plofker explains:

> The universe is created and destroyed during one *kalpa* or day-and-night period of Brahman, which lasts for 4,320,000,000 years. There is a shorter period called a *mahayuga*, or 'great age', of 4,320,000 years: it is divided into four smaller intervals in a 4 : 3 : 2 : 1 ratio, during the course of which the world decays from good to bad, as in the Golden, Silver, Bronze and Iron Ages of Greek legend. The last and worst of these sub-periods is the *Kaliyuga*, which is one-tenth of a *mahayuga*, or 432,000 years long.

These numbers of years may have been large, but the numbers denoting cosmic distances were still larger; one measure of the size of the heavens ran to 18,712,069,200,000,000 *yojanas* (a yojana being a unit of distance in the region of twelve kilometres). To insert such vast counts into verse in their verbal forms could hardly be satisfactory, creating a need for a special notation for numbers. Several different systems were tried, testimony to the difficulty of the problem and the urgency with which a solution was needed. Most proved to be brief experiments that hardly outlived their inventors, but one would become the world's most widespread number notation.

One system, for instance, used by the author Aryabhata in the sixth century, assigned consonants to small numbers and vowels to powers of 10 in order to represent numbers as syllables: nonsense syllables, admittedly, but capable of being read aloud. The idea did not catch on, but another scholar – Haridatta – devised a variant in the following century, in which the thirty-three Sanskrit consonants were each assigned to one of the digits from 1 to 9. That meant there were several possible choices for each number; the four choices for '1' were *k, t, p* and *y*, and the system is therefore known as *katapayadi*. This was a positional system: the first letter written denoted the units, the next the tens, the next the hundreds, and so on; and the position was the only indication of whether a given symbol represented units, tens or hundreds. This system became important in number magic and divination, and remained in use for centuries, although it never became the dominant system for writing numbers.

A different system again assigned one series of phonetic signs to the whole numbers up to 9, another to the multiples of ten up to 90, and sometimes also another to the multiples of 100: this system had the same structure as the Egyptian demotic and the later Greek alphabetic number symbols. It became the usual way to label the page numbers in books, surviving in some regions as late as the nineteenth century.

A different option, and one that avoided strings of nonsense syllables, was to create verbal synonyms for number words: such as 'earth' (meaning one); 'eye' (meaning two); 'sense' (meaning five); and so on. Like the *katapayadi* numbers, these could be strung together starting with the units, followed by the tens and the hundreds, as many times as was needed in order to express even the largest numbers in the Sanskrit cosmology. This had the advantage of involving actual words and being therefore much better suited to insertion into verse meant to be learned by heart. The earliest examples appeared in astrological texts in the third century CE, and this scheme, too, would remain popular for several centuries before fading away.

So much for the less successful experiments – there were more – illustrating what a way of writing numbers down could look like. But of course, the Brahmi script also possessed a set of dedicated number symbols, which made their first fateful appearance in inscriptions of the third century BCE and spread through the subcontinent during the following century or so. In this earliest form, there were distinct symbols for the numbers from 1 to 9 and for the multiples of ten from 10 to 90. Much has been written of the rival possibilities that they were independently invented in India or imported from elsewhere, or at least inspired by a foreign model. The structure of the system in this archaic form was once again the same as the Egyptian demotic one, and about a third of the signs show some resemblance to their Egyptian equivalents. Traders from the Ptolemaic kingdom certainly reached the Malabar coast in this period, and conversely Buddhist missionaries travelled to Alexandria in the third century BCE. So the possibility of influence is real.

The Brahmi number symbols were used in inscriptions, on coins, on land grants inscribed on copper plates, and to number the lines of verse, and they spread into central and south-east Asia by the sixth or seventh century. From those drier climates, manuscripts on wood, palm leaf and paper survive, in which the Brahmi numerals are used within texts in the Indo-European language called Tocharian. In southern India they continued in use until the eighteenth century; in Tamil, numbers with this structure continue in use today.

These symbols were not used for arithmetic or accounting, however; in mathematical contexts they did not displace the use of number words or the other systems of symbols described above. Not, that is, until an innovation occurred in their structure.

The idea, simple enough, was that instead of using the full set of symbols, just those for the numbers from 1 to 9 would be employed, with the convention that (starting from the right) the first number denoted units, the next tens, the next hundreds and so on. It was a positional system on the same lines as the *katapa-yadi* one. And if there were, for instance, no tens to record, the blank place was marked with a dot.

It is not certain when the innovation first occurred. Dates as early as the mid-first millennium BCE have been suggested on the basis of a possible link with the symbols used on Chinese counting boards. But the earliest undisputed direct evidence for the Indian place-value system occurs in inscriptions from the late seventh century CE, both in Southeast Asia and in India itself. The moment of innovation must lie somewhere in between.

The idea of a positional system, with a special sign for a blank column, spread throughout the region; by perhaps 800 CE, a preference for it was widespread. The actual shapes of the symbols continued to evolve in tandem with the complex family of different writing scripts used in South Asia: a process that continues to this day, when the major South Asian languages have their own alphabets and number symbols. But all structure their number symbols using the same decimal place-value system inherited from their Brahmi ancestor.

One of the tasks for which this system was well suited was written calculation, something that on the whole is unusual in the history of numbers and counting. Most cultures calculate using devices – like the counting board or the abacus – supplemented by mental arithmetic for simpler operations. But, as most readers of this book will know only too well, a place-value system of number symbols enables operations like addition, subtraction, multiplication and division to be done on paper using algorithms that a young child can learn, as long as the child is willing to memorise the addition and multiplication tables, each with a few dozen entries. It therefore tends to turn both counting and calculation into special kinds of writing, and to bring them into contact with a child's early training in wielding writing implements and surfaces: to make learning and manipulating the number symbols a natural accompaniment to learning the letters and arranging them into words and sentences.

———

Oh Lilavati, intelligent girl, if you understand addition and subtraction, tell me the sum of the amounts two, five, thirty-two, one hundred ninety-three, eighteen, ten, and a hundred, as well as the remainder of those when subtracted from ten thousand.

The famous Indian mathematician Bhaskara II was one of the many heirs to the rich tradition of writing words and numbers in India. He lived in the twelfth century, in Vijjadavida, a city in the western region sometimes called the Great Escarpment of India or the Western Ghats, the north–south range that separates the Deccan Plateau from the coastal plain beside the Arabian Sea. A richly forested biome of exceptional biodiversity, it was at this time part of the Western Chalukya Empire, ruled at the time of Bhaskara's birth (1114 CE) by the successful Vikramaditya VI; though Bhaskara's lifetime saw the start of that empire's breakup as Vikramaditya's feudatories fought their rulers and each other. His was an agricultural society, with well-organised guilds and a complex taxation bureaucracy. Trade linked it with China, with Southeast Asia and with the Baghdad caliphates to the west.

Literature and scholarship flourished under the Chalukya kings, who patronised scholars writing in both the local Kannada language and in Sanskrit. Bhaskara's ancestors were said to be court scholars: astronomers and astrologers. It is also said that he learned astronomy from his father, Mahesvara.

Bhaskara himself may have worked as a court astronomer, but very little is known in detail about his life. His writings, however, are among the best known of the whole Sanskrit mathematical tradition. By his day, the sections in astronomical texts dealing with arithmetic had become long and complex and had started to appear as independent treatises, called *ganitas*, setting out the art of calculation as a subject in its own right. Writing in this tradition, he produced his *Lilavati* – literally 'the beautiful' or 'the playful'. It has long been rumoured that the book was addressed to his own daughter, though there is no positive proof one way or the other, and within the text the addressee is sometimes imagined as masculine, sometimes feminine.

The *Lilavati* discussed arithmetic comprehensively. Introducing the numbers themselves, Bhaskara named seventeen different powers of ten: 1, 10, 100, 1,000, *ayuta, laksa, prayuta, kota, arbuda, abja, kharva, nikharva, mahapadma, sanku, jaladhi, antua, madhya, parardha*. He explained how to write numbers down: in his verses, numbers were given in words, but in the prose sections they were written using the place-value system.

From the place-value system, the *Lilavati* continued through the operations of addition, subtraction, multiplication and division, and on to finding squares and square roots, cubes and cube roots, followed by operations involving fractions or zero. It concluded with algorithms to solve what would now be called algebra problems, rules for reasoning about quantities in proportion to each other, and some geometry. The arithmetic alone filled dozens of Sanskrit verses, and the prose sections which provided examples for each rule expanded its length many times over. The examples ranged from the pedestrian to the picturesque:

Tell me, quick-eyed girl, if you know the correct procedure for inversion, the number which, multiplied by three, added to three-fourths of the result, divided by seven, diminished by one-third of the result, multiplied by itself, decreased by fifty-two, having its square-root taken, increased by eight, and divided by ten, produces two.

A traveler on a pilgrimage gave one-half of his money at Prayaga, two-ninths of the rest at Kast, one-fourth of the remainder in toll fees, and six-tenths of the remainder at Gaya. Sixty-three gold coins were left over, and he returned with that to his own home. Tell me the initial amount of his money.

There is a hole at the foot of a pillar nine *hastas* high, and a pet peacock standing on top of it. Seeing a snake returning to the hole at a distance from the pillar equal to three times its height, the peacock descends upon it slantwise. Say quickly, at how many *karas* from the hole does the meeting of their two paths occur?

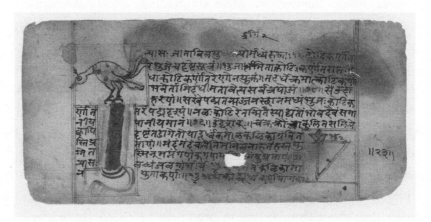

The 'Peacock' problem, in a manuscript of the *Lilavati*.

No very early manuscript of the *Lilavati* survives today, but chance survivals of other materials from the period show what it would have looked like. Palm leaves – or in northern India, birch bark – were used as a writing surface, and books traditionally had a rectangular format, broader than they were tall. The leaves were kept together not by binding along one edge but by passing a string through a hole in their centre. Boards and a cloth wrapper could be used to protect the whole volume, but even so it is rare for a manuscript of this kind to survive more than a few centuries at most.

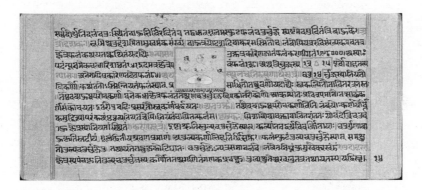

Brahmi number symbols set off from the text in a box,
in a manuscript of the *Lilavati*.

The writing itself formed a rectangular block of text in the centre of each page, without capital letters and with only a few spaces between words: there were no line breaks between verses, nor any page breaks between chapters of text, though sometimes red was used to highlight the starts of chapters. One of the few interruptions to the continuous stream of the text, in fact, was that the many number symbols were often set off with a box around them.

In 1207 Bhaskara's grandson set up, with funds from a local ruler, a school for the study of his writings, which included as well as the *Lilavati* a book on algebra and two on astronomy. It is from the foundation inscription of that school, at a temple in Maharashtra, that most of the little information about Bhaskara's life derives. His works became the standard and best-known mathematics textbooks in the Sanskrit tradition, with many thousands of manuscripts in circulation. He may justly be credited with doing more than anyone else to promote the decimal place-value system as a tool for writing numbers and for the practice of written calculation.

Ibn Mun'im: Dust numerals

Marrakesh, in North Africa: between the mountains and the sea. Around 1200 CE.

Take a bundle of silk threads and make a tassel. Take silk of two colours and make a brighter tassel. Take three colours. Take ten.

Or: start with ten different colours and choose only six of them for your tassel. How many ways are there to make the choice? The mind leaps ahead to the abstract questions: how does the number of choices change as you alter the overall number of colours or the number chosen for the tassel? And what other kinds of choice are governed by the same numbers? The number of letters taken from the alphabet to make a word, perhaps, or the number of books selected from a shelf, the arrangement of stresses in a line of verse . . .

Bhaskara did not know it, but by his day the Brahmi number symbols were already well advanced on the journey that would take them to practically every country in the world. Ways of counting can spread fast in the right circumstances, and widely. Systems of number symbols, perhaps particularly so. They are not tied to any one language, so they can easily cross cultural borders that might be insurmountable for a method of writing words.

Number symbols are easier to learn than verbal scripts, more transparent, more self-explanatory. As evidence for this, the numerals in several scripts can be read today even though their words remain undeciphered: the 'Linear A' of Minoan Crete, the enigmatic writing of the Indus Valley civilisation.

Buddhist travellers took some information about Indian astronomy to India's neighbours – China, central Asia, Iran – in the first half of the first millennium. In 662, the Christian bishop Sebokht at Nisibis on the Euphrates remarked on the 'subtle discoveries' of the Hindus in astronomy: their methods of calculation and their use of 'nine signs' for computation. He may or may not have known about the tenth sign, the dot or circle marking an empty place.

Later, the spread of Islam brought the vast Arabic-speaking world into direct contact with India; Muslim merchants reached the Malabar coast by the eighth century. In the 770s, an Indian mission brought a text about astronomy to the court of al-Mansur in Baghdad. Such texts routinely included expositions both of the Brahmi number symbols and of basic arithmetic using them, and made use of the symbols in their extensive astronomical tables. The symbols were becoming visible to Arabic speakers during the eighth century, even if no exact moment of 'first contact' can be identified.

The culture which the Brahmi number symbols now entered already possessed several sophisticated ways of counting. Counters were present in the form of coinage, whether Byzantine gold or the dinars first minted in 697 by the Umayyad caliphs. The Arabic language inherited a decimally based set of number words from its proto-Semitic ancestor, which were therefore ultimately related to those of Akkadian and Hebrew as well as to those of the more distant linguistic cousin Egyptian: *wahid, ithnan, thalatha, arba'a, khamsa*. A system using the letters of the Arabic alphabet to stand for numbers, structured exactly like that of the similar Egyptian and Greek systems, existed, and indeed continued to exist beside the Brahmi-derived symbols right up to the fourteenth century.

The Roman system of hand-signs for numbers up to 9,999 was also known. They were sometimes described as 'Rum', meaning Roman; the term here denoted the eastern half of the Roman Empire, Byzantium. The hand-signs were widely used to store intermediate values when carrying out arithmetic mentally, and down to the tenth century some writers in Arabic continued to recommend their use, a system 'which calls for no materials and which a man can use without any instrument apart from one of his limbs'.

Despite all this, though, the Indian number symbols had their supporters from early on, and they evidently found much success. A great movement to import and translate texts from Persian, Greek, Syriac and Sanskrit took place in the eighth and ninth centuries under the Abbasid dynasty, centred on Baghdad. Texts were copied, studied, taught from, discussed, improved and commented upon. Learned mathematicians made great strides in astronomy, algebra, trigonometry and other fields. Some experts took the names of Greek scholars as epithets, like al-Uqlidisi, 'the Euclidean'. Around 825 CE, Muhammad ibn Musa al-Khwarizmi wrote in Arabic a 'book on Indian calculation', in which he described the place-value system and its use in calculation, and advocated it for its usefulness and simplicity. Al-Khwarizmi also revised or translated a set of the famously number-heavy Indian astronomical tables, possibly those that had been originally brought to Baghdad fifty years before.

Over the following decades and centuries, there grew up a whole genre of Arabic writings on the Indian system of number symbols and its calculation procedures. All started by describing the nine symbols and noting that a small circle was used to mark a space where there was no symbol. They followed with chapters on the various operations: addition, subtraction, multiplication, division; then on fractions and perhaps the sexagesimal system (inherited

via Greek from cuneiform) used in astronomy. Further topics were
the finding of square and cube roots, for which procedures existed
broadly similar to that for long division.

The processes that were described, promoted and taken up,
however, were at first subtly different from their Indian models.
Several of the early Arabic texts used in their titles or their descriptions the terms 'board' (*takht*, originally a Persian word) or
particularly in the west of the Arabic world 'dust' (*ghubar*). The
idea was that a scribe would carry a board; when it was to be used,
he would strew sand on it and write the number symbols in the
sand with a finger or stylus. This made it easy to erase, replace and
move symbols as the calculation required. The whole practice came
to be called in Arabic 'dust arithmetic' or 'board arithmetic', the
number symbols themselves 'dust symbols'. This made arithmetic
a wholly temporary process, leaving no permanent record the way
the Indian procedures on paper and bark did.

Paper was certainly known in the Islamic world from the
mid-eighth century onwards (the story is that Chinese prisoners
taught its manufacture around the year 750), and already by the
following century some authors were advocating its use for calculation in preference to the dust board. The mathematician al-Uqlidisi
argued that the use of the dust board demeaned the serious scholar
because of its association with 'misbehaved' people 'who earn their
living by astrology in the streets', and that its difficulties included
dirty fingers and sand being blown away by the wind. The key
argument, though, was that using paper produced a record of the
process of calculation, so that if you needed to check or find an
error you could do so: unlike the *tahkt*, on which at least some
steps of the calculation were necessarily erased along the way.

Whether on board or on paper, it is likely that the number
symbols quickly began to spread from their initial context of learned
astronomical calculation and tabulation, into the arithmetic taught
to children and used by merchants and other practitioners. By the

later ninth century, and more so through the tenth and eleventh, references to the new notation can be found across the writings of Arabic historians, encyclopedists and religious scholars. Eventually, a mathematical tradition extended over the whole of the Arabic and Islamic world, from northern India to the Atlantic coast of North Africa and Iberia.

The far west of the Islamic world was itself an important centre of culture and learning. One of its exponents was Ahmad ibn Mun'im al-'Abdari, born probably in the second half of the twelfth century, and so roughly a contemporary of Bhaskara 8,000 kilometres to the east. His origin was in the town of Denia in al-Andalus, on the coast near Valencia. At this time the Almohad dynasty controlled both Islamic Iberia and the Maghrib on the North African coast. Theirs was a reforming, purifying impulse and its leaders proclaimed themselves the true Caliphs. Controlling vast resources – mines for sulphur, iron, lead, mercury and cinnabar, textile and ceramic industries – and the whole of the trans-Saharan trade as well as trading connections north of the Mediterranean into Italy, they supported and promoted culture from architecture and poetry to theology, mysticism and philosophy. The Jewish philosophers Averroes and Maimonides both lived under Almohad rule. Medicine, mathematics and astronomy flourished.

Most of ibn Mun'im's life was spent in the Almohad capital at Marrakesh, where he studied both medicine and mathematics. He was reckoned one of the leading specialists in geometry and number theory, and wrote on both Euclidean geometry and the arrangements of whole numbers known as magic squares.

His book 'On the Science of Calculation' was one of those that described the Indian numerals and their use. Aimed presumably at students, it covered the usual topics: the shapes of the 'dust numerals' (still so called, though now routinely written on paper); addition, subtraction, multiplication, division, extraction of roots and fractions.

Midway through the book came a section of more advanced discussions. These dealt with topics like the sums of series of odd numbers, of even numbers, of square numbers or of cubes. Ibn Mun'im discussed triangular numbers and other similar arrangements, and the so-called perfect numbers, which are equal to the sum of their factors. And he examined at some length the question of how many words can be made from a given set of letters.

Questions of this general type had been asked for centuries, by linguists and philosophers interested in enumerating the different roots of Arabic words under various constraints. As early as the 790s, scholars performed laborious enumerations of two-, three-, four- and five-letter roots. Up to the twelfth century, such problems were mainly approached simply by listing possibilities and counting them one by one. One writer constructed and described a device consisting of a disc with two rotating wheels, on each of which were written the letters of the Arabic alphabet. By rotating and aligning the wheels it was possible to find all the different alignments of three letters more quickly.

Ibn Mun'im's approach was mathematical rather than mechanical, and he was keen to calculate rather than simply count. He worked through a series of subsidiary problems on the way to his answer to the general question; in fact he began by considering an ostensibly different problem. Given ten colours of silk, with which you wish to make tassels, how many different choices can be made of, say, five colours? Or three? Or seven? In other words, how many ways are there to choose a certain number of colours out of ten, if the order in which they are chosen does not matter?

There are obviously ten ways to choose just one colour. By simple counting, you can discover that there are forty-five ways to choose two, a hundred and twenty to choose three . . . Ibn Mun'im made a table showing the different ways to choose, all the way from choosing one colour from a selection of one, to choosing ten from a selection of ten. By observing the properties of the table, he was able to see ways to calculate its numbers rather than finding them by laborious counting. The table was identical to the arrangement of numbers later called Pascal's triangle; each number was equal to the sum of the number above it and the number to the left of it.

The ways to choose a set of letters from the twenty-eight available in Arabic are naturally the same as the ways to choose a set of colours from a selection of twenty-eight. To choose three letters (or colours), for instance, there are 3,276 possibilities. This was just the start, though, and ibn Mun'im worked through a dazzling series of more complex questions and examples. Suppose one letter is repeated: how does the number of possible choices change? Suppose more than one letter is repeated. And so on. One of his computations asked, in effect, how many distinct words of nine letters can be formed by rearranging the string ABCCDDEEE. Finally, he introduced restrictions on the alternation of vowels and consonants in order to respect the way in which pronounceable words were really formed in Arabic. For several of these cases, ibn Mun'im constructed separate tables showing the answers – often enormous numbers – arising from a combination of calculation and plain counting.

———

Empires come and go. Muhammad al-Nasir (1199–1213), at whose court ibn Mun'im worked, is probably best known to history as the leader who lost the battle of Las Navas de Tolosa in 1212 to a coalition of Christian kings, a watershed for the fortunes of the Almohads and the beginning of that empire's collapse in both Iberia and Africa. He died a year later, and internal divisions hastened the end of the Almohad dynasty and the disintegration of its empire later in the century, to be replaced by several smaller Islamic monarchies.

The Arabic learned tradition continued, of course, and ibn Mun'im's results about combinations and permutations were passed on to succeeding generations. The subject illustrates a pervasive tendency of advanced exercises in counting to shade into calculation, tabulation and eventually algebra; and, more generally, for the Indian number symbols and calculation on paper to facilitate thinking about numbers in ever more abstract ways.

The question of exactly what the Indian number symbols looked like in the Arabic world is a surprisingly elusive one. The fact is that contemporary manuscripts of the earliest Arabic works on numbers and arithmetic have not survived. For al-Khwarizmi's work on Indian-style calculation, indeed, even the original text is lost, and the only evidence of what he wrote consists of later translations into Latin. There is absolutely no direct evidence for what the number symbols looked like to him. For ibn Mun'im's work on arithmetic and combinatorics there is only a single manuscript, made generations after the original author. In fact, the fragility of writing materials has swept away all of the first three centuries of evidence for the Indian number symbols in Arabic contexts.

When the evidence does fade in, during the eleventh century, in the form of manuscripts, astronomical tables and even astronomical instruments such as astrolabes, there is – not surprisingly – a good deal of variation from manuscript to manuscript, reflecting slightly different conventions at different places and times. In fact al-Uqlidisi, as well as advocating calculation on paper using the decimal place-value system, proposed replacing the unfamiliar Indian symbols with the first nine letters of either the Greek or the Arabic alphabet. But that idea did not catch on, and the Indian symbols were retained, albeit modified in detail.

One curious change took place throughout the Arabic-speaking world. In the Brahmi script, which is read from left to right, numbers were written with higher powers on the left. That is, the eye would come to the higher powers first (as it does in modern languages using the Latin alphabet). In Arabic, which is read from right to left, the order of the numerals was *not* reversed to match the direction of reading: the higher powers still appeared on the left. That meant that the eye would reach the smallest power, the units, first.

As to the shape of the individual number symbols, the tradition eventually split into two branches: the symbols of the eastern and western halves of the Arabic-speaking world, both clearly descended from the original Brahmi forms used in Sanskrit. A Moroccan writer of the late twelfth century remarked on the difference between the two styles, but the divergence had certainly arisen long before that. The Baghdad-centred area, where al-Khwarizmi and al-Uqlidisi had worked, came to use a set of symbols that are the ancestors of those used in modern Arabic: ١, ٢, ٣, ٤, ٥, ٦, ٧, ٨, ٩. The western regions of the Arabic world, however, had a different style of calligraphy and differences in the form of certain letters. All the number symbols were modified to one degree or another there, and the symbols for 6, 7 and 8 became quite different from their eastern counterparts. This western style of writing the Arabic numerals – sometimes called the Maghribi numerals – persisted until the eighteenth or even the nineteenth century before finally being generally replaced by the eastern system.

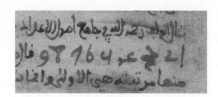

Indian number symbols in the western Arabic form,
from a manuscript made in fifteenth-century Granada.

The interest of the western-style Arabic number symbols lies in their onward journey: for it was this form that would travel to Europe and become the ancestor of the 'Arabic numerals' known around the world today. Indeed, during the lifetime of Bhaskara and ibn Mun'im, that process was already under way.

Hugo of Lerchenfeld: Toledan numerals

Similarly to the admiring reference by the Bishop Sebokht in Syria in the seventh century, the Arabic numerals received a first glimmering mention in Europe in the form of a passing reference in a manuscript about something else. The monk Vigila, of the monastery of Albelda in the Rioja, was responsible, adding to his copy of a Latin encyclopedia the following words:

> the Indians have a most subtle understanding and all other peoples yield to them in arithmetic and geometry and the other liberal arts. And this is clear in the nine figures with which they are able to designate each and every degree of each order [of numbers]; of which these are the shapes . . .

The shapes given were essentially those of the western-style Arabic numerals. This was in 976, the period of the Umayyad caliphate based at Cordoba: evidently some of the vast learning of that splendid city had made its way north into Catalonia. And thus, paradoxically, the earliest surviving evidence for the western forms of the Arabic number symbols comes from manuscripts written in Latin.

During much the same period around 970, a young man named Gerbert of Aurillac travelled to Catalonia, there to study mathematics for three years. Once back in France, he would become archbishop of Rheims, and subsequently Pope, as Sylvester II. At Rheims he devised and promoted a new – some would say bizarre – kind of counting board, whose key feature was that the counters were marked with number symbols. Instead of putting, say, six counters in the tens column, you would put a single counter there: one marked with a '6'.

For two centuries, Gerbert's board was the preferred tool for learning number lore in the monastic schools, with around three dozen separate treatises devoted to it. One or two actual 'boards' in this style – made of paper – survive to this day, but sadly none of Gerbert's specially marked counters. In the later treatises, the symbols on the counters were Arabic numerals, and it is likely that that was Gerbert's own choice, deriving from his sojourn in Catalonia and what he had learned there.

The procedures for adding, subtracting, multiplying and dividing on Gerbert's counting board were essentially identical to those for written number symbols on a dust board, the counters being successively replaced by others as the computation progressed. This, then, was to all intents and purposes a European version of the Indian – and now Arabic – practice of calculation using written symbols. Instead of writing the symbols on a surface, you placed them, moved them and removed them in the form of the counters. A minor difference was that on a board there was no compelling need to mark an empty column with a 'zero' counter: you could just leave it empty. In fact, 'zero' counters do seem to have existed, redundant though they arguably were. In Latin the zero was at first sometimes called a 'wheel' and sometimes a *theta*, after the Greek letter whose shape it resembled. But fairly soon, Latin writers settled on terms derived from the Arabic *sifr*, meaning 'null' or 'void', itself borrowed from the Sanskrit *sunya* meaning 'empty'.

In Latin the word became *cifra* or *chifra*, the ancestor of words in modern European languages such as English 'cipher' and French *chiffre*. In these languages, the term sometimes came to designate not just the zero but the whole set of number symbols, or indeed their use ('ciphering'). Sometimes the term's meaning was extended to symbolic characters in general, especially secret ones. Thus in Italian something '*cifrato*' is in a secret code, in French a '*chiffre*' may be a monogram initial letter, in German '*Ziffer*' means the system of Arabic numerals, while one of the meanings of 'cipher' in English is simply a 'zero'.

Gerbert's counting board was cumbersome to use compared with more traditional models, and there is little evidence that it was used outside the teaching context of monasteries and their schools; little sign that it was used for real-world calculation rather than for learning about numbers and their manipulation in the abstract. After the later twelfth century, there is no further evidence for it even in the monasteries, and it seems to have fallen out of use and out of memory.

Meanwhile, and more widely in Europe, Arabic numerals had begun to appear among the manuscripts entering the continent through Antioch, through Sicily, and increasingly through Spain. Copies went as far north and west as France and England by the mid-twelfth century. They prompted a period of experiments, perhaps broadly similar to the Indian experiments in representing numbers in writing: most short-lived, and some very strange.

Some European scholars tried to make a hybrid of Roman and Arabic number symbols, by using a selection of the Arabic symbols as abbreviations within a system structured like the Roman one. The results were monsters like *MC87* for 1,187 or *52M,CC20* for 52,220. Others took up the place-value system from the Arabic numerals, but not the Arabic symbols. Abraham ibn Ezra, in Pisa, used the first nine letters of the Hebrew alphabet in a system that was structurally identical to the Brahmi numerals; a colleague in Flanders used the first nine letters of the Latin alphabet in the same way. Evidently the new number symbols were stimulating and provocative, but equally evidently they created confusion and anxiety just as often as clarity.

It was in the context of astronomy and astronomical tables that European writers first learned to use the Arabic numerals properly. The Spanish route was becoming the most important one for Arabic material entering Latin Europe, and the city of Toledo was established as a major centre of translation and transmission. The astronomical tables of al-Khwarizmi received adaptations for use first in Cordoba and later in Toledo. The Toledan version acquired a wide reach and importance in Latin; more than two hundred manuscript copies survive today, and these 'Toledan tables' were in turn adapted for use in various other European cities, becoming a model for other astronomical works in Latin.

This material was of interest in Europe because it could potentially solve urgent problems with the Latin calendar. The underlying difficulties were to do with the fact that there are not a whole number of days in a year: that is, for each orbit it makes around the sun, the Earth spins on its axis not 365 or 366 times but a fractional number between the two. The problems would not receive a lasting solution until the Gregorian calendar reform of the late sixteenth century, which produced the rules about leap years still in use today.

Absorbing Arabic calendar lore, these scholars also learned how to use the Arabic numerals. The Toledan tables were accompanied in some manuscripts by Latin versions of al-Khwarizmi's book about the numerals, which explained their use for arithmetic. The association of the numerals with Toledan material was so strong that some Latin commentators called them the 'Toledan' – rather than the Arabic or the Indian – numerals.

When Arabic (or Toledan) numerals began to appear in Latin manuscripts, they once again underwent changes in their exact shapes, though they remained clearly derived from ancestors of the western Arabic type. It is possible they were influenced by Visigothic conventions for writing and abbreviating Roman numerals, but the differences may just be the result of the usual vicissitudes of handwriting style over several centuries. At least two Latin writers actually remarked on the difference between the 'Toledan figures' and the original Arabic ones, giving examples of each, although it was understood that the place-value system in which they worked was the same; that the differences were superficial.

One unlikely centre of expertise and experiment in both astronomy and the Toledan numerals was the German city of Regensburg. There, a small circle of interested scholars collected together a clutch of mathematical manuscripts, including copies of the Toledan tables and of a Latin version of al-Khwarizmi's arithmetic book. They tried out different ways of adapting the thirty-year Arabic calendar cycle, or the nineteen-year Jewish one, to Christian use. They even indulged in distinctive local variations on the shapes of the number symbols, giving '2' and '3' long vertical tails and occasionally writing '3' to look like a sort of cleft stick.

It was here that, so far as the evidence goes, Arabic numerals first began to leak out of the contexts of astronomy and the study of calendars, into more everyday uses of numbers and counting.

Hugo, citizen of Regensburg, became a canon at the cathedral there after the death of his wife, probably in 1178, and remained so until his own death in or after 1216. He is an obscure figure, his life recorded as the merest outline: except that he spent a proportion of his time compiling an idiosyncratic notebook. Now in the Bavarian State Library in Munich, it contains seventy-eight sheets of parchment, kept together not by binding but by wrapping them in a larger piece of parchment.

It was a messy compilation, constructed in part by copying – with variations – a volume of annals from the neighbouring monastery, and by supplementing it after its end in 1167. As Hugo grew older, his handwriting became larger and messier in the later entries. Other parts of the book consisted of inserted passages and pages with notes on all kinds of subjects: family history, letters, the Hebrew alphabet, spiritual reflections, solar and lunar eclipses . . . The loosely inserted sheets must have been in constant danger of ending up dropped, lost or disordered.

Altogether, the annals ran from 35 to 1201 CE by the time Hugo ceased to write. He evidently also had astronomical or arithmetical material from the monastery in front of him, because instead of using Roman numerals for the dates in his annals he used the Arabic numerals: not MCLXVII but 1167; not MCCI but 1201.

It is hard to convey just how eccentric Hugo was to date his annals in a way that now seems like the natural choice. No more than a handful of his contemporaries in the city would even have recognised the Indian/Arabic/Toledan symbols as numbers, never mind been able to read them. Perhaps in a private notebook, it didn't matter; perhaps Hugo relished the novelty, the cleverness, the hint of secrecy about the new numerical code. Thus he became in his small way a pioneer, the first in Europe to use the Arabic numerals outside the realm of astronomical tables or arithmetical treatises.

A page from Hugo's notebook, showing the Toledan number symbols.

Bhaskara, ibn Mun'im and Hugo of Lerchenfeld were alive at the same time, in the late twelfth century. They wrote down numerals in their Sanskrit, Arabic and Latin texts within a decade or so of one another. Their stories tell of the enormous reach the Indian number symbols had attained by this date. In India they already had a history of hundreds of years; of four centuries in the Arabic-speaking world. In Latin contexts, their use on parchment was only just beginning – though they had been manipulated on counters since about 1000 – and only hindsight makes it clear that they had a future in the Latin world, and that that future would involve the Toledan version of the numerals rather than one of the other experiments or versions of the period. Chance played a role in this, as did the high status of Toledo as a centre of translation from Arabic, and the prestige and visibility of the texts issuing from that city.

The Indian numerals had, of course, a vast subsequent history in all three locations. In the Latin world, at least four different versions of al-Khwarizmi's treatise on Indian-style arithmetic had been made by the end of the twelfth century, representing various degrees of reworking and refinement as translators attempted to make the material more comprehensible to their readers. If the place-value system itself was familiar from the counting board – whether traditional or Gerbertian – the idea of calculation on parchment seems to have required laboured explanation, demonstration and discussion. So widespread did these texts become that the name 'al-Khwarizmi', which in Arabic meant the man from Khwarezm, came to mean the process of calculation with the Arabic numerals, or a treatise describing it. The Latin word, inevitably somewhat garbled, was *alchorismus* or later *algorismus*. Eventually it came to mean arithmetic in general, whatever the method, and in modern languages an 'algorithm' can be any step-by-step procedure. It was quickly forgotten that the word had its origin in a personal name, and some later medieval authors even attempted to reason back from *algorismus* to 'Algus' or 'Argus', its supposed inventor. Chaucer once referred without explanation to 'Argus, the noble counter'.

The numerals in the Toledan form gradually increased in visibility and importance; in country after country a critical mass of the literate switched to using them for calculating, and for recording the results of counts and calculations. Their proponents effectively captured the narrative and succeeded in associating with the Arabic numerals a sense of prestige that went rather beyond their practical advantages. The older system – counters for counting and calculation, and Roman numerals or words for communicating results – had worked for a thousand years and had enabled the Romans to administer their vast empire: it was not in any clear sense broken.

Italy took up the new way first, with both merchants' and public accounts adopting the Arabic numerals alongside the techniques of double-entry bookkeeping between the thirteenth century and the fifteenth. Specialist schools taught the techniques; by the early 1500s Florence, for instance, had six schools teaching up to twelve hundred pupils 'algorism'. Northern and Western Europe followed, with the proportion of Arabic numerals in documents Europe-wide hitting 50 per cent by around 1500. (There appears to be no evidence for the claim, still occasionally repeated, that the new numerals – or zero in particular – were resisted, feared or blanket-banned because of their foreign, Hindu or Muslim associations or some supposed association with magic.)

Hindsight has given this rise a feeling of inevitability, as though the claims of late medieval boosters in favour of the numerals were the unvarnished truth. They were indeed efficient for calcu-lation, if you were prepared to invest the initial effort in memorising addition and multiplication tables, and learning algo-rithms for addition, subtraction and so on. And, as practised on paper rather than dust boards, slates or blackboards, they provided a record of the process of calculation, making it relatively easy to check and spot errors. But boards and counters continued to be important well into the era of print, tally sticks were used at the English exchequer until late in the eighteenth century, and Roman numerals linger to this day in contexts where prestige or stability need to be communicated (the numbering of monarchs and popes, the dates of films and public foundations) or ambiguity between two series of numbers needs to be avoided (page numbers in books, month and day numbers in certain styles of date).

What did counting in Europe look like, after the Arabic numerals had completed their conquest?

The account keeper: Counting on paper

A Dutch interior in the 1650s. A wooden counter: inkwells, a ledger.
A few coins. A middle-aged woman sits at her books, the window
open to catch the light. She is working on her accounts: perhaps
for her household, perhaps for her business. Beside her, keys, another
reminder of her responsibilities. A barrel, some jugs, a bowl; a
basket with the hint of food or drink in it. It is a cluttered room;
it could even be the corner of a shop or a kitchen. From the wall
behind her, a bust of Juno peers down: patroness of commerce,
she is a sort of guardian of the woman's goods and wealth. Beside
Juno, a huge map larger than the counter and the accountant
herself: Africa, East Asia, the Americas; dragons and serpents
beyond.

Sadly, the account keeper has fallen asleep. Though her pen is
still poised over the ledger – the top one of a stack of three ledgers
in fact – her eyes are closed, her head slumped into her hand. The
perils of overwork; the weight and tedium of calculation with
number symbols on paper.

Nicolaes Maes, who painted *The Account Keeper* in the Netherlands
in 1656, was one of Rembrandt's most talented pupils, and had a
long career as a portraitist as well as producing a few dozen works

like this one depicting Dutch interiors. Most were closely observed studies of women at work: making lace, peeling vegetables, praying, reading – or doing the accounts. Maes's own craftsmanship was exquisite, reminiscent of Vermeer in its control of colour and light. With *The Account Keeper*, he created a rare record of what the practice of accounting looked and felt like in the Dutch golden age.

Nicolaes Maes, 'The Account Keeper'.

The painting only shows a tantalising glimpse of the accounts book itself: but accounts books survive in huge numbers from the sixteenth century to the nineteenth, deriving from households, businesses and individuals. Each page could be as simple as a list of

items and their prices, with a total at the foot of the page (and some
working by the side, providing a record of how that total had been
arrived at). Other books were elaborately ruled in multiple colours,
with separate columns for profits and losses, and the 'total' column
updated after each entry: much like a modern bank statement, and
manifesting the same awkward combination of dullness and importance. If something is not right here, it can mean that something is
terribly wrong – fraud, theft, embezzlement – in the real world of
counters and goods. Maes's accounts book looks like the more
complex kind, and the fact that there are two more volumes stacked
closed on the counter suggests a business or household of some
complexity, in which different categories of income and expenditure
have to be tracked separately, or in which a single run of accounts
has to be compared over a long period of time.

How did you learn to do such things? For this, too, a great deal
of evidence has survived. Now that counting was well established
as part of reading, and calculating as part of writing, you learned
it wherever you learned your reading and writing. At home or at
school, you made a 'cyphering book', in which you wrote out
definitions and exercises starting with 'numeration' and passing
through addition and subtraction, multiplication and division until
you – possibly – moved on to more advanced procedures like
reasoning about proportions or keeping accounts. The word
'ciphering' came from 'cipher': that is ultimately from Sanksrit
sunya and Arabic *sifr*: a little reminder of the long heritage of this
way of recording numbers, and this curriculum of study.

Hundreds of the books survive, from all over Europe and North
America. They typically consisted of a quire of paper sewn together,
in a generous size: often 30cm by 20cm. Children from well-to-do
families might buy one ready-made, but most students would have
sewn, roughly bound and decorated the book themselves. Some
ruled lines to help them write: these were meant to be fair-copy
books, but the students using them began as young as nine or ten

and often had quite uncertain handwriting. Many, all the same, turned out careful and neat, with different inks elaborately used, headings, borders and careful management of the space on the page to make a truly personal textbook. Illustrations were common, ranging from the accomplished watercolour to the childish doodle.

The basic technique of instruction was always the same. The teacher would provide a 'rule' for some arithmetical process – numeration, addition or subtraction, say – which the student would copy into the book. The teacher would also provide some examples of the use of that rule, to try out as practice on slate or waste paper. The students, called forward individually, would show their exercises. If they were wrong, they had to be done again; once they were right they could be copied into the ciphering book.

Estimates varied, even at the time, as to how well this worked. Sometimes wrong sums found their way into the ciphering book if the teacher was careless; sometimes students got the answers right merely by copying from one another. And sometimes work that was correct but used a slightly different method from the teacher's was judged wrong unfairly. It could be a frustrating system, and it was certainly a repetitious and a laborious one. A student who stayed in school for several years might produce several hundred pages of ciphering work; some students filled more than a dozen separate books. The ciphering system assumed that counting was part of reading and writing, and that the way to learn arithmetical operations was by repetitiously writing them down: that the route to the mind was through the pen and the hand. There was no mental arithmetic, and precious little speaking at all.

The curriculum started with 'numeration': learning to read and write the Arabic numerals fluently. Most students wrote out a table showing how to read Arabic numerals up to perhaps eight or ten digits long: hundred millions, ten millions, millions . . . all the way down to hundreds, tens and units. After that, practice: write eighty-five in numbers. Write one hundred and eight. Write one million, eleven hundred and one. You would do these on scrap paper until you or your teacher thought they were right, then copy them into the book.

Often there was an emphasis on large numbers. Up to nine digits was typical, but some textbooks went up to twenty or more; one went to seventy-eight digits, the perhaps ludicrous 'duodecillion'. Large numbers provided more arithmetical work per exercise, of course; they also gestured none-too-subtly at prosperity, motivating students – or trying to – by promising that if they learned this material well they too could one day deal in hundreds of cows, thousands of barrels or millions of silver coins.

Following numeration came the four arithmetical operations: addition, subtraction, multiplication and division. The syllabus, its order, and even some of the details of the algorithms for calculation were the same as they had been in medieval Europe, in the Maghrib and Baghdad, in India itself. Next, the 'rule of three', which taught how to find the missing number from a set of four linked quantities. If it takes three men two days to build a certain wall, how long will it take five men to build the same wall? The same subject, with some strikingly similar examples, was covered in Bhaskara's textbook.

Once a student had done all that, she unfortunately had to start all over again with so-called 'denominate' numbers: that is, numbers denoting money or weight or measure. The systems in use were numerous; just for weight there were troy weight, avoirdupois weight and apothecaries' weight, and there were many more for volumes, lengths, areas and of course money. All of this taught the arithmetic that would enable you to keep domestic or small-business accounts in adulthood, dealing with the costs of food, clothing and equipment as well as bargains, barter and loans:

Divide twelve pounds, 17 shillings and six pence equally among four men.

Bought a hogshead of molasses for twenty-one pounds and five shillings. How many gallons did it contain, valuing the gallon at two shillings and six pence?

Two merchants A and B agree to trade together. A puts in £332 for four months, and B £475 for seven months; they gained £520. What was each merchant's part of the profits?

Many ciphering books tailed off rather than finishing neatly: students left school, and the series of definitions and examples came to a more or less abrupt halt. Nevertheless, adults frequently retained their ciphering books as personal textbooks, using them to display their competence at arithmetic and handwriting to schools, colleges or potential employers, or referring to them for the techniques of calculation that were useful in daily life. Some added to them, modified or corrected the original exercises. Some books were passed down among siblings or from parent to child; some became dense palimpsests recording the arithmetical education of a whole family. However orderly the sequence of definitions and exercises, chaos had a tendency to creep in around the edges.

Maes's *Account Keeper* is a tranquil scene, invested in the details of wood and plaster, crockery and curling paper. It says something about the dignity of household work, and of accounting in particular. It speaks of the dedication it required, the time and devotion to be given to the task. Of an activity worth pursuing to the point of exhaustion. While the woman's clothing is that of a solid, not a wealthy, householder, three separate ledgers would be a lot for a merely domestic set of accounts. The map on the wall behind her seems a clear hint that her ledgers dealt with trade beyond the borders of the Netherlands: perhaps far beyond, to the ends of the world she knew.

Did the artist intend to moralise about the woman and her labours? The picture has points in common with traditional depictions of avarice or worldliness: the prominent scatter of coins; the prominent image of, literally, the world. It would not take much to turn it into a depiction of a miser in a counting house: the resulting accounts book might look almost identical, but it would be designed to provoke a rather different response and of course tell a rather different story about what counting and the practices derived from it do to people.

Yet this woman is not a miser, and the fact that she has fallen asleep makes her look just as much like an emblem of melancholy, perhaps even of sloth. The ledgers are about to fall to the floor, the orderly scene to collapse (the woman to wake, disturbed by the crash). It was Maes's style to look on such things with a gently humorous eye: his other pictures of people asleep invite a laugh or at least a wry smile, as dreamers have their pockets picked or risk a scolding for neglecting their chores. Here, Juno herself – the guardian of goods and wealth – also seems to have nodded off, echoing the sleepiness of the account keeper, and hinting that matters are not as well stewarded as they seem. And the world map, the sign of the woman's ambitions and international connections, is half a reproduction of contemporary maps, half a fantasy, embellished with fantastic beasts copied from maps of the stars.

Keeping accounts – writing numbers down – always comes with a tinge of anxiety that the stability and solvency they record might one day collapse. It always has the character of an island of order and rule, threatened on every side by chaos. It was the character of the Renaissance and the Enlightenment to cultivate and expand those islands of order, bringing more and more aspects of the world under the purview of numbers, tables and facts. But a glimpse of chaos often remains at the edges.

Caroline Molesworth: Counting the weather

Monday, 1 January 1825. Cobham, Surrey. After a night of wind and rain, it is a mild day for the time of year, with the thermometer in the vestibule standing at nearly twelve degrees by nine in the morning. Quantities of furze are in bloom. Thirty-year-old Caroline Molesworth takes a clean, bound notebook and writes the first of her year's series of weather observations.

What could be more chaotic than the weather? Yet what more tempting for the project of gathering data, tabulating, averaging and reducing that chaos to order?

One of the uses of the Arabic numerals, from a very early stage, had been the making of tables and lists: activities which of course predated the numerals themselves and had appeared in – for instance – astronomical work in ancient Mesopotamia. Lists of goods, their quantities and prices; lists of observed phenomena such as the positions of moon and planets; lists of calculated quantities such as trigonometrical ratios or predictions of planetary positions. In terms of sheer quantity, astronomical tables of the 'Toledan' family probably account for the majority of written number symbols in later medieval Europe. The compactness of the

Arabic numerals was a help in these contexts, and their place-value feature perhaps did something to facilitate perusing the data once it was tabulated: if all the units line up vertically, all the tens will line up as well, and so will all the hundreds, making the numbers easier to read, their patterns easier to spot at a glance.

Astronomical tables began to appear in print in the later fifteenth century, and they were quickly accompanied by other kinds of printed tables: calendars, arithmetical ready reckoners, tables to work out the interest due on loans, tables to convert weights and measures from one system to another. The printed almanac became a best-selling forum for simple printed lists of all such types, typically combining a calendar with astronomical predictions in numerical form while also providing useful information of the ready-reckoner kind: rates of interest, the conversion of currency, the equivalence of local and foreign weights and measures. Johannes Gutenberg himself printed an almanac, and Europe and North America saw millions, even tens of millions printed during the long height of their popularity through the eighteenth and nineteenth centuries.

Almanacs and astronomical tables of this sort generally contained, in the main, predictions based on theory. But tables of data as actually observed also acquired a new range and visibility in the world of print, contributing to the long rise of a statistical mindset in European and North American populations through the eighteenth century and beyond. Francis Bacon, to whom modern empirical science looked as founder, had urged the collecting and tabulating of observations – the gathering of facts – as a key activity in the creation of knowledge: effectively as a way of understanding the world in its own right. Under the impetus of scientific societies like the Royal Society of London from the later seventeenth century, such harvesting of facts became increasingly widespread, a favoured activity of amateurs wishing to contribute their mite to the store of knowledge. Many of the facts thus gathered were – or could be made – numerical, the outcomes of counting or measuring the natural world; their publication in many cases amounted to tables of number symbols. Tables of mortality, for instance: the numbers of people dying each week from various causes. Tables of population, of national wealth, of government spending. Tables of patients successfully or unsuccessfully treated, of success and failure in controlled experiments.

Indeed, counting successes and failures in experimental trials was a crucial step in the development of the idea that by observing the frequency of events and outcomes you can make reliable predictions about what will happen in the future: statistical inference, in other words, one of the foundations of modern science. Enumerating events in this sense was surely the most important innovation the Enlightenment brought to counting.

Weather predictions, for example, had long appeared in printed almanacs. Weather observation was promoted by the learned societies and became a widespread enthusiasm from the late seventeenth century, a form of what would now be called citizen science. During the eighteenth and nineteenth centuries, weather journals were kept across Europe, from Ireland to Portugal to Czechoslovakia, in China and Japan, around the USA and the Caribbean, in New Zealand and Australia. Their authors included clerics and schoolteachers, merchants, landowners, physicians and gardeners, whose motivations ranged from the study of climate and its effects on human health to anxiety about crop performance and consistency in planting and harvesting times.

By the 1720s, the Royal Society alone was receiving dozens of weather journals from around Britain, Europe and North America, to the point that there was more data to hand than could be either printed or synthesised, and many remained unpublished and unused in the Society's archive. Other British publications such as the *Gentleman's Magazine* carried accounts of the weather from various contributors, and the tradition was taken up by local societies of natural history who maintained it well into the nineteenth century. There were even book-length publications of the weather records of particular observers or locations.

The unsystematic character of both observations and publications became a matter of comment, and institutions occasionally attempted to impose more order and routine on the collecting of weather data. Robert Hooke, a founding fellow of the Royal Society, had published a specimen format for weather journals as early as 1667. Enterprising printers offered blank weather-report forms for sale. A favoured model was the ship's log, which was subject – at least in national navies – to a literally military discipline and typically recorded a consistent set of information including eventually the air's temperature and pressure as well as more general observations, generally at regular times of day and night.

—

Caroline Molesworth, born in England in 1794, was descended both from the Baronets of Pencarrow and from French immigrants, and after a childhood spent in Cornwall and London she moved with her mother to Cobham Lodge, in Surrey. There, in October 1823, she began the series of weather observations she would continue for the rest of her life. She collected and cultivated rare plants, and one of her reasons for an interest in the weather was its effect on her garden.

Molesworth acquired – perhaps to some degree sought – a reputation for benign eccentricity: 'brusqueness and originality', in the words of one commentator, as well as 'good sense and feeling'. She retained for the rest of her life the style of dress current around 1800. 'Very kind to the poor and generous to her relatives',

she was described by those who knew her well as a 'most enter-
taining companion'.

She acquired various instruments to aid her observations of the
weather. Outdoor thermometers for the maximum and minimum
temperatures, and two more thermometers for different positions
inside the house. A rain gauge made in London, which amounted
to a bottle sunk in the ground with a funnel in its top. A barometer by the well-known instrument maker George Adams. She also
had a 'storm glass': a bubble of glass containing air and water,
whose appearance responded to both temperature and pressure.
The reputation of this scientific toy as an instrument for serious
observation was never high, and in April 1843 Molesworth silently
stopped reporting its behaviour in her weather journal. The house
must also have had a wind-vane, but nowhere in her observations
did she say so, merely reporting the direction of the wind (never
its speed). For some, this collection of instruments would have
been an object of pride, even of conspicuous consumption: but
for Molesworth they were instruments of use far more than of
display.

Her journals became slightly battered over their years of service,
with covers of marbled board and yellowed pages occasionally
spotted with ink or – perhaps – rainwater. Superficially, they looked
very like a set of household accounts. The pages were ruled in red
and blue, and Molesworth recorded up to nineteen columns of
data each day, filling a two-page spread twice a month: the day
and date, the times of sunrise and sunset, the phase of the moon and
the hour(s) of observation. Temperatures at various locations,
atmospheric pressure, cloud cover, wind direction, inches of rainfall, general observations about the weather, about the plants and
animals in her garden.

The discipline of daily and twice-daily observations was evidently
congenial to Molesworth. In a sense, it became part of her persona.
Where some diarists defined themselves through their acute study

of human nature, of political affairs or of their own minds and bodies, she used her neatly ruled pages to record a self that was pure scientific observer. Travel, illness or accident she reported only when they affected the series of weather records.

A typical page from Caroline Molesworth's weather diary.

The series continued in all for over forty years. After about 1850, though, the gaps due to illness became more frequent, and other hands than Molesworth's filled in an increasing proportion of the entries. The latter volumes of the series gradually petered out, the last full observation being made early in October 1867. A final, terse note came on Wednesday, 9 October 1867: 'rain'.

Molesworth's tables of data and of number symbols were the outcome of a variety of processes: observation, measurement, counting and calculation. At the end of each year, she prepared summaries which relied particularly on both computation and counting. At the fullest, there was a summary for each month of the year, for which she counted the number of days on which rain had fallen and the number of days on which the wind had stood in each of eight directions, as well as computing the month's total rainfall and the mean and ranges for the barometer and three thermometers. There was also a summary for each whole year, for which Molesworth provided a similar set of totals and means. On occasion she even prepared multi-year summaries. Francis Bacon would have been proud.

Molesworth did not just collect data; she also corresponded with a number of other weather observers and meteorologists, owned and read books on the subject, and published in periodicals some of her summaries of the weather of Cobham. On her death, her collection of dried plants was presented to the herbarium at the Royal Gardens at Kew, and her weather diaries were presented, at her request, to the Meteorological Society in London. In 1880, a volume of summaries extracted from them was published. Their editor picked up on the statistical side of the work, writing that

> Miss Molesworth's labours will not have been . . . useless if they add anything to the amount of information which we may look forward to from the careful observations now being carried on in the same field of research, promising – it is not too much to say – to be of infinite value to the country agriculturally, by showing the bearing of weather influences on the growth of our food crops.

Caroline Molesworth's work illustrated both the pleasures and the frustrations of systematically observing the natural world. On numerous occasions, observations had to be reported as missing or questionable because of breakages or other difficulties among her instruments. Every trip away from home occasioned a gap in the series of observations that could never be made up or substituted for. And the very process of dividing winds into eight points of the compass, rounding temperatures and pressures to the nearest whole number, and so on, emphasised that turning observations into numbers must always involve simplification and, therefore, loss.

The process of aggregation at the end of each year possessed something of the same ambiguity. On the one hand, it promised to transform a series of isolated observations into something bigger; a series of atoms into a coherent picture. On the other hand, though, the summing and averaging effected a real loss of texture and granularity compared with the original observations. And even on the scale of a whole year, the result could still feel like a series of details rather than a description of a climate. In 1827, Caroline Molesworth counted one hundred and forty-one days with rain or snow or frost, twenty-three of them in December; forty-two days of northerly wind, eleven of them in September. There were fifty-six days of southerly wind, spread across every month of the year. The most common wind over the year was from the northwest . . . It fell to her posthumous editor to note certain relationships between the dates at which plants flowered and variations of temperature and rainfall from year to year. Perhaps it would not have been possible to turn the numbers into a bigger picture during Molesworth's lifetime.

————

By the nineteenth century, and still more by the twentieth, Arabic numerals were everywhere, carried to most of the countries in the world, their family tree now dense with branches. Generations of textbooks along the lines of Bhaskara's and Al-Khwarizmi's presented 'numeration' as the reading, writing and visualising of Arabic numerals on dust boards, blackboards, parchment or paper. It had become common in ordinary speech and writing to say 'numbers' and mean the Arabic number symbols, as though the two were the same: as though there were no conceivable or at least worthwhile other way of representing numbers. (The neat way to expose the paradox here is to observe that many people would say 2,543 is a four-digit number, but no one would say Mary is a four-letter girl. Arabic numerals become identified with the thing they represent far more readily than words do.)

Meanwhile, the number concept itself was being steadily expanded beyond the 'natural' numbers possible with beads, fingers or tallies. Fractions and ratios had been written down in the ancient Near East; irrational numbers like certain square roots – which cannot be expressed by any fraction – were already of interest in ancient Greece. The many writers of ciphering books studied negative numbers, and Molesworth used them to record temperatures. In the context of mathematical research, 'real', 'imaginary' and 'complex' numbers would be defined during the nineteenth century, and in the twentieth, 'hyperreal' and 'surreal' numbers too. Much of this was supported by the power and flexibility of the Arabic number symbols, steadily extended using negative signs, decimal points and other devices.

This may seem like an end point; it may seem, even, as though the rise of the Arabic numerals is *the* story of counting. But it is in reality only one branch among many. Writing numbers down as symbols is only one of many ways of counting; and indeed the distinctive structure of the Arabic numerals is only one of several ways a set of number symbols can be organised.

INTERLUDE

Number symbols

As the other examples in this book illustrate, counting symbols have displayed a range of different systems at different times and places. The basic choice is between having one symbol repeated several times (|, ||, ||| . . .) and having a set of different symbols to symbolise the different numbers (α, β, γ, . . .). The first system is that of the simplest tallies, whether ancient or modern; the second that by which the books of Homer's *Iliad* and Aristotle's *Metaphysics* are numbered to this day. But as with counting words – and number symbols very often imitate their structure from the counting words in the language of their first users, albeit frequently with some tidying up – a system with no more structure than this quickly becomes unusable for larger counts.

If a number base is used – say, a special symbol for the number ten – it can be combined with the smaller numbers simply by juxtaposing it: XI meaning ten-and-one, for instance. For multiples of the base, the choice must then be made again, between simply repeating its symbol the required number of times, and adopting a whole series of symbols to denote the different multiples. In Roman numerals, for instance, ten, twenty and thirty may be shown by X, XX and XXX; in Greek by I, K and Λ. Another alternative is to reuse the original set of number symbols with some sort of modification: *a*, *b*, *c* becoming *a'*, *b'*, *c'*, say, to show that they now mean ten, twenty and thirty. Or, finally, you can make no modification, but rely on the relative position of the symbols to show that some mean units and others mean multiples of the base. This last is the system of the Brahmi numerals that became the dust numerals, the Toledan numerals, and finally the Arabic numerals.

Each of these ways of structuring a system of number symbols has been used over long periods; each is capable of remaining stable over hundreds or even thousands of years: none has in fact any very strong tendency to evolve into – or be replaced by – any of the others. The advantages and disadvantages of different kinds of system depend very much on what you want to use it for. Some are quicker to learn, some quicker to write; some use a smaller set of symbols overall, others tend to represent any given number with a shorter string of symbols. Says Stephen Chrisomalis, historian of number symbols, 'There is no ideal numerical notation system; rather, each system is shaped by a set of goals that its users and inventors seek to attain, and that they can achieve only by compromising on other factors.'

About a hundred different sets of number symbols have appeared throughout history; from the dawn of writing until about 1500 CE, their number steadily increased. But number symbols – like alphabets – have a marked tendency to be reused by more than one culture; the successful ones have been particularly successful at travelling, perhaps particularly successful as tools for communication between different communities, cultures, languages and places. The cuneiform numbers, the Egyptian demotic and Greek systems, the Roman numerals, and of course the Brahmi numerals all travelled widely, as did many more. The Arabic numerals were eventually carried around the world with – mainly – the European languages and cultures; other systems meanwhile became extinct. In the last five hundred years the number of distinct systems has fallen, on the whole.

As recently as the 1990s, it was not uncommon to hear it suggested that the decimal place-value system effectively represented the end of history as far as number representations, counting and arithmetic were concerned, no significant improvement being likely or even possible. Yet such certainties are no sooner stated than they begin to crumble. There have always been other ways to count, and there will always be other systems of number symbols. And there was an element of chance in the meteoric rise of the Arabic numerals: place-value systems have been invented at other times and places without overrunning the world. For the specific purpose of doing arithmetic in writing, the Arabic numerals have real advantages: but spare a moment to consider how long it takes most children to learn the forty-five addition facts and thirty-six multiplication facts in the base-10 tables, and how long to become fluent at even a limited set of operations on larger numbers: addition, subtraction, multiplication and division. (The extraction of roots, present in many medieval and Renaissance textbooks, seems to have largely dropped out of view in modern classrooms.) How easy it is to make a mistake, after all, despite the vaunted advantages of the system: and what a relief it is nowadays to outsource the whole thing to a calculator. Similar effort devoted to analogous tasks with different ways of representing numbers – the counting board, say – can make them, too, feel efficient and seem natural. Chrisomalis again: 'We do not stand at the end point of a linear historical sequence, but in the midst of a branching and complex yet patterned and explainable world of written numbers.'

So, not an end point after all; and at most times and places the history of counting has not been written at all, but has involved other ways of keeping track. The drift of the last century has been that way even in the parts of the world dominated by Arabic numerals. In Molesworth's own lifetime, there came the first hints that the era of performing large tabulations, calculations and even counts by hand, using Arabic numerals on paper, would not last for ever. Machines were beginning to appear that could perform some of those functions automatically, and whose descendants would eventually transform the way humans related to counts, numbers and data. Over the last few decades, the prevalence of number representations in the world has increased exponentially, but the overwhelming majority of them are not Arabic numerals on paper but something quite different: binary representations encoded in electrical impulses.

Another hugely important branch on the tree of counting must surely be the story of physical devices – manual, mechanical and eventually electrical and electronic – that assist with the counting process: the story, in other words, of counting machines. Like the Arabic numerals, they have swept the world, making their story into a global one; but nowhere is their deep history more richly recorded than in East Asia.

6

Machines that count: Around East Asia

From African roots, the story of counting spread its branches through the Fertile Crescent, Europe and India. East Asia is home to another branch, one that has undergone its own evolution for millennia. Here the number words are also decimal for the most part, as in the large language families further west. Here number symbols are powerfully in play even in some of the earliest extant texts. But here counting also involves a distinctive set of counting devices. Everybody has seen an abacus – the *suanpan* in Mandarin, the *soroban* to the Japanese and the *jupan* to Koreans – whose operators became famous far beyond East Asia for the skill and speed with which they worked.

After the abacus, electrical machines. Right up to the 1950s it was a matter of comment that they worked more slowly than the most skilled abacus operators. But eventually the machines overtook the humans, and it is now in the form of electrical and magnetic signals that most of the world's counting is done: the binary representations deep inside modern digital devices. This is a global story, with early technical innovations coming from the USA in particular. Today, East Asia is a leader in both the manufacture and the consumer uptake of the products that make the digital revolution.

And before the abacus? Before the abacus, East Asia was home to a different technology for recording and manipulating counts, whose operators also attained astounding dexterity and efficiency. It was in use for centuries, and the patterns it made provided the shapes of the classical Chinese number symbols. That technology was the counting rods.

Hong Gongshou: Counting with rods

Zhili, Qimen district, Huizhou prefecture, China. The fourteenth year of the Chongzhen emperor (1641 CE). Hong Gongshou is being assessed for tax liability.

He reports to the assessor his own age (forty-seven) and the ages of his brother, son and wife. The assessment document also records the size of his fields, from a survey carried out a decade earlier, and their liability for the summer and autumn taxes. It goes into detail about the size of eight plots of land including a paddy field and a patch of marsh, and it closes with a note that the family's house is constructed of straw and possesses three rooms.

Hong's summer tax is given as 6 *dou* 4 *sheng* 4 *ge* 1 *shao* of wheat; the autumn tax 1 *shi* 6 *dou* 3 *sheng* 9 *ge* of husked rice.

The same scene was played out millions of times each year across Ming China, as tax assessors tried to get a grip on who owed what under the country's complex system of assessments. Revenues were raised across a land area of 4 million square kilometres, from the Pacific to the Himalayas, and from a population of more than 150 million; the total collected amounted to perhaps 5 or 10 per cent of the country's agricultural output.

For administrative purposes, the vast country was divided into thirteen provinces, then into prefectures, sub-prefectures and counties. At the lowest level, groups of 110 households were organised into *li*, each under a rotating *lichang* (chief, alderman) who performed, among other things, the collection of taxes. The emperor controlled it all from the capital at Beijing, in a system which relied heavily – at least in theory – on his direct control. Some of the questions brought to his attention were very small: 'the relocation of a particular business tax station, or from which productive area a particular county was to draw its salt supply, or how many rolls of silk were to be awarded to a foreign tributary mission'.

Such matters were passed upwards from the village level and assembled by a pyramid of officials in the prefectures and provinces, headed by the minister of revenue. Those same structures also organised the immense task of moving the goods – after collection – across the country to where they were distributed. For instance,

as late as 1578 the imperial university in Nanking still received 3,500 *piculs* of husked rice from Ch'ang-chou prefecture, 100 *piculs* of wheat from Ning-kuo prefecture and 100 *piculs* of green lentils from Ying-t'ien prefecture, all in South Chihli, and also over 20,000 *catties* of dried fish from Hukwang province, all these items being charged to the tax quotas of the respective districts.

As well as direct levies on agricultural output, there was an obligation to provide labour services and various specialised kinds of goods – office supplies, military equipment, medical supplies, even brooms to sweep the palace – from time to time. Furthermore, the assessments were adjusted based on a ten-yearly survey of land and an annual census of the number of people and amount of property in each household. Though the system was simple in theory, it inevitably tended to burgeon into increasing complexity, with surcharges and additions to cover the transport of the goods, and an ever-increasing willingness to substitute silver bullion for the set of items nominally due. This led to complex calculations, with outcomes that sometimes had to be specified in thousandths of an ounce of silver. A modern commentator suggests that even the simplest cases might each have taken an hour and a half to calculate, giving the example of 'a labor service payment of 0.0147445814487 taels of silver on each *picul* of grain in basic land tax assessment', a situation which 'provided a paradise for lower-echelon tax collectors and bookkeepers'. Attempts at consolidation and simplification were made in the sixteenth century, but they were only ever implemented patchily.

Even if there was a tendency for some of this to become a mere formality, with survey and census numbers being carried over from one period to the next unchanged or only mechanically updated, nevertheless the administrative effort involved was huge, as was the volume of tax records being generated. By Hong Gongshou's time some were complaining that the whole exercise was a waste of paper and writing brushes. Four copies of each tax assessment were made: one each for the county, prefecture, provincial and imperial governments. The last was the subject of an elaborate storage regime on an island on lake Hsuan Wu near Nanjing. By the 1640s there were 1.7 million volumes in seven hundred storage rooms, very probably the largest archive the world had yet seen.

The largest archive, and therefore the largest collection of number symbols. Chinese, then as now, had multiple ways of recording the outcome of a count or a calculation, and the surveys and censuses certainly reflected some of that variety.

One way to write a number was simply to write the logograph for its word. These already appeared in effectively their modern form in inscriptions of the third century BCE, and continue in use to this day. The system, both spoken and written, is a decimal one, with words for the numbers from one to nine and for 'ten', 'hundred', 'thousand' and 'ten thousand' which combine as, for instance, *two-myriad four-thousand three-hundred six-ten five* (24,365). A zero was introduced to the written system from the thirteenth century, if not earlier, but strictly speaking it is redundant. The words 'ten thousand', 'thousand', 'hundred' and 'ten' already specify what each unit sign means: there is no need for its position to do so as well.

This system spread throughout East Asia, appearing in Japan by the third century and Korea by the fifteenth. Calligraphic variants of the signs – which did not change the structure of the system – began to appear from the first century BCE. At least two such variants, the so-called accountants' numerals and the secret numerals, are still in use today, their greater visual complexity providing a guard against forgery.

But what technology was used actually to count and calculate in China and its neighbours? Texts from the first millennium BCE mention sticks being used to count, and by at least the third century BCE they were employed in a system that involved laying them out in groups on a flat surface. The sticks were never completely standardised, and could be made of bamboo, wood, ivory, bone or other materials. In the classical period they ranged between about 9 centimetres and 14 in length and were perhaps 7 millimetres wide: 271 of them could be made into a hexagonal bundle and held in one hand.

The patterns in which the sticks were arranged functioned as number symbols. Counting up to five was done with rods laid down vertically beside one another, just like the most primitive tallies. Numbers from six to nine were shown by the combination of one horizontal with one, two, three or four vertical rods (once again, the eye never needed to deal with a single group larger than four). The same patterns were repeated for the tens, hundreds and so on, with the larger powers placed to the left. The interesting complication was added that for the even-numbered digits – tens, thousands, hundred thousands and so on – the pattern of sticks was rotated by ninety degrees, to help distinguish the digits from one another. A rhyme reported that:

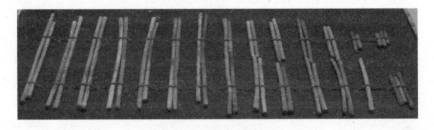

Chinese counting rods

Units are vertical, tens are horizontal,
Hundreds stand, thousands lie down;
Thus thousands and tens look the same,
Ten thousands and hundreds look alike.

Any number of any size – subject to the limits of how much flat space you had and how many rods you possessed – could be denoted in this way.

Literary references seem to place the sticks on tables, beds or the floor. But for the system to work really efficiently, the surface on which they were placed probably needed to be marked in some way: otherwise the gap between one group of rods and the next would have the potential to become ambiguous, 24 looking all too similar to 2004. Large wooden boards would have been cumbersome, and none have turned up as archaeological artefacts. But there is at least one reference to a cloth marked with columns or rows for the sticks to occupy.

Like the Chinese counting words and their equivalent symbols, the counting rods spread to Japan and Korea, and they were much used throughout East Asia for 2,000 years. They were employed not just for counting, but also for calculation. Algorithms for the basic arithmetical operations in the rod system are not at all difficult to devise, and eyewitness reports suggest that they could be performed very rapidly; one expert in the eleventh century 'could move his counting-rods as if they were flying, so quickly that the eye could not follow their movements before the result was obtained'. Both mental and written calculation seem to have become unusual, displaced by the more efficient rods. By the twelfth century, perhaps earlier, algorithms had been devised not only to extract square and cube roots using the counting rods, but to find numbers satisfying systems of relationships that a Greek would have represented using the geometry of squares and rectangles, a modern European using the algebra of linear and quadratic equations.

The counting rods also prompted their users to write down numbers in the same form: to make sets of ink strokes that matched the layout of the rods. As early as 400 BCE, Chinese coins bore numbers using symbols that seem to be of this kind; and like the rods themselves, these number symbols remained in use until the Ming period. It has sometimes been suggested that this set of number representations in fact travelled as far west as India and was an inspiration for the place-value structure of the Brahmi number symbols. Direct evidence is lacking, and well-informed experts disagree about how plausible the suggestion is. It is certain that by the eighth century CE, the Brahmi numerals themselves had been seen by some in China, and were the subject of comment that compared them unfavourably to the counting rods as a technique of calculation: 'The method is complicated . . . and you are lucky if you get it right.'

The rods and their techniques are constantly assumed – and in some cases described in detail – throughout the rich tradition of

Chinese mathematical writing: in the set of canonical teaching texts codified in the seventh century, in later learned writings and also in collections of more practical problems. One such collection, compiled in the thirteenth century, ranged across practical subjects such as chronology, surveying, architecture, military problems and trade. Among its sections was one dealing with taxes and service levies of precisely the kind to which Hong Gongshou and millions like him were subject. The problems give a vivid sense of the difficulties taxpayers and tax collectors faced throughout the Ming period. Farmer A makes over 407 *mou* of paddy fields to farmer B and 516 *mou* to farmer C; how will this affect the three men's tax liabilities, in rice and silk? Land in a certain prefecture has emerged from the sea. How should taxes be apportioned among the six villages that claim parts of it across nine different kinds of field? Twelve thousand men are levied to transport military rations to the frontier, in proportion to their field tax liability and in inverse proportion to their distance from the frontier: how many men will have to come from each village?

That last question was to gain a new topicality in the early seventeenth century, as the Ming regime faced the series of crises that led to its downfall. From 1618 onwards there were Manchu invasions; Chinese military spending spiralled and taxes rose commensurately. But the system was not able to mobilise enough of the land's vast resources with sufficient speed. Although revenues quintupled in just twenty years, the army could not reliably be paid.

To add to the disaster, northern China also saw poor harvests in the mid-1630s, compromising the possibility of raising the sums that were needed, and ultimately triggering local rebellions. Flood, famine and disease arrived to complete the picture. In fact, the year 1641–2 was the last in which the tax survey was drawn up. The Ming state collapsed soon afterwards, in 1644.

The vast collection of documents held in Nanjing was destroyed when the city fell to the Qing army in 1645. One report holds that this, the world's largest collection of written number symbols, was used as kindling for gunpowder in the last defence of the city. Documents like Hong Gongshou's tax return are rare survivals today, reminders of a system long since swept away. For the counting rods and the number symbols based on them were also destined for oblivion. Both for counting and for calculation, another device was now on the rise: the *suanpan*.

Kiyoshi Matsuzaki (and Thomas Wood): Counting with beads

Nobody knows when it was invented, or indeed where. The *suanpan* in Mandarin, the *soroban* to the Japanese and the *jupan* to Koreans. It seems to have been during the sixteenth and seventeenth centuries that the *suanpan* took over in widespread use from the counting rods as a way of representing and working with numbers, though it made its first appearance at least as early as the thirteenth century, and perhaps earlier still. The idea, like all great ones, is simple enough: a rectangular frame holds a set of parallel wires on which beads are strung. The position of the beads represents a number, and the rapidity with which a few fingers can shift the beads to change or calculate with the number is the evident reason for its success. Its operators became famous far beyond East Asia for the skill and speed with which they worked.

Its name says something about what it is, and indeed about what mathematics is. Historian Joseph Dauben explains:

The character 'Suan' is based on a radical meaning cowry for shells or, in a slightly different form, goods. . . . 'Suan' originally referred to the counting board, with which bamboo counting rods were used to carry out calculations during transactions, and later came to refer to the abacus or any method of calculation, including mathematics generally. Thus the character 'Suan' ideogrammatically embodies both the original objects and methods that became the stock in trade of the court and administrative mathematicians.

The frame is usually of wood, often of bamboo; but a *suanpan* can be made from many materials. The number of rods never seems to have been standardised, but there is a noticeable tendency over time for larger numbers of rods to be envisaged and for them to become concomitantly smaller, so that the whole device always remained of portable or at least manageable size. By the twentieth century, a typical *suanpan* for commercial use might leave only a millimetre of clearance between one column of beads and the next, and allow little more than a centimetre of free wire for the sliding of the beads.

Like its predecessor the counting rods, the *suanpan* represents numbers using a decimal system. Which row represents the number 1 is a matter of choice, depending on the need of the moment; once it has been chosen, its neighbour to the left will represent the tens, the next row the hundreds, and so on. In fact, a beam running across the board divides each wire into an upper and lower section: the beads in the lower section are worth 1, 10, 100 and so on, those in the upper 5, 50, 500 . . . Beads are brought into play by being pushed against the bar.

The oldest Chinese *suanpan* had two 'fives' beads and five 'ones' beads, the second 'fives' bead enabling numbers larger than 10 to be briefly represented on a single wire during calculation. In Japan, the 'fives' beads were reduced to a single one during the second half of the nineteenth century, and the 'ones' beads to four a few decades later, producing the most streamlined decimal *soroban* possible with just five beads in total on each wire. This Japanese type has since become widespread in Korea although it has – particularly in China – not wholly supplanted the older types.

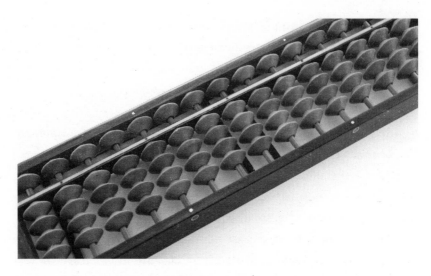

A Japanese *soroban.*

It is remarkable that the modern Japanese *soroban* therefore has the same structure of four-beads-plus-one as some ancient Greek and Roman counting boards. If transmission occurred in either direction – perhaps during the Middle Ages – it seems to have left no direct evidence, and this may instead be a case of two cultures converging on similar solutions to analogous problems of calculation and efficiency. The 'abacus' that began to appear in Europe and the USA from the nineteenth century is usually traced to Russian models brought to France during the Napoleonic period; but the ancestry of the Russian *schoty*, which has ten beads per rod, is itself obscure. The fact that the English word 'abacus' can refer to any of these devices – Greek, Roman, medieval, Russian or East Asian – only adds to the confusion.

Some very detailed descriptions of efficient finger technique for the *suanpan* and *soroban* exist. The most minimal version has the 'fives' bead (or beads) moved up and down by the index finger, while the 'ones' beads are moved up by the thumb and down by the index finger. But three-fingered or even two-handed techniques also appear in some books. Arithmetic with the *soroban* proceeds from left to right, from larger place values to smaller. The addition and multiplication tables are of the same size and complexity as those that have to be learned for calculation with Arabic numerals, or indeed with any decimal representation of numbers. The techniques of arithmetic are closely analogous to those with the number rods, which represent numbers using a similar structure of units, fives, tens, fifties, and so on. At one time, it was also usual to memorise a division table, based ultimately on counting-rod techniques: but it was phased out in the first half of the twentieth century in favour of a method of division that used the multiplication table only. Just as with the rods, it is perfectly possible to do more advanced arithmetic such as the extraction of square and cube roots using a *soroban.*

The *soroban* continued to gather interest and expertise, through refinements to the device itself and to its techniques, right up to the end of the twentieth century if not beyond. Its visibility and its operators' skill probably reached their apex in the second half of that century. By that time it was frequent for children to learn the use of the *soroban* at home, even before entering school. In both Japan and China, *soroban/suanpan* techniques were part of the formal curriculum from typically third or fourth grade. Vocational schools preparing students to go into business similarly provided intensive *soroban* training. And for those wishing to become experts, there were extracurricular programmes usually associated with primary schools, with training that might occupy several hours per week. Patient practice and memorisation could produce results that were impressive by any standard, and national institutes organised exams and competitions: in Japan from the 1930s, and more recently in China. The levels of skill on display on these occasions were startling: in the Japanese system, where most examinees were teenagers, even the lowest grade of examination involved the addition of five columns of fifteen multi-digit numbers in five minutes, with 70 per cent accuracy required. At national level a contestant could expect to add or subtract fifteen numbers of up to nine digits in little more than a minute.

———

Tokyo, Monday 12 November 1946. The US Army's Ernie Pyle theatre. An audience of nearly three thousand GIs assembles to witness such classics of entertainment as 759.843 × 57.941 or 4,768,788,098 ÷ 14,593, performed on a pair of calculating machines. According to reliable witnesses, they cheer frenziedly.

By the nineteenth century, it was already a matter of comment among Western observers that *soroban* calculation was strikingly faster than paper-based alternatives. In the context of the American occupation of Japan after the Second World War, the *soroban* was famously put to the test against one of the new electric calculating machines. *Stars and Stripes* magazine organised a head-to-head contest between a US Army private operating a machine made in California, and a Japanese *soroban* expert.

Barely a year after the end of the war, the Japanese economy and infrastructure remained shattered, the country occupied by foreign troops. Half of Tokyo itself had been destroyed by Allied bombing. The event was planned as an entertainment for American troops, and reports suggest that an easy victory for the electrical machine was expected. There were to be two rounds of addition problems, three each of subtraction, multiplication and division, and finally a round of more complex composite problems requiring several operations. The numbers ranged in size up to twelve digits.

Private Thomas Wood was chosen following an internal competition, and appears to have been the best operator of the electric calculating machine to be found in Japan at the time. He worked in the finance disbursing section of General MacArthur's headquarters. Kiyoshi Matsuzaki, his opponent, was a clerk in the communications ministry; a famously skilled operator of the *soroban*, he was known to the American sponsors as 'The Hands'. Both contestants were aged twenty-two. Nearly 3,000 GIs packed the venue, and many of them placed bets (on Wood, presumably).

In fact, Matsuzaki won four of the five events, and overall eight of the twelve rounds of problems. He completed the addition problems in around 1 minute 15 seconds each time, and the others in little more. In one case, Wood trailed more than thirty seconds behind him: thirty seconds that must have felt more than a little excruciating for Wood and his fans, cheering no longer.

The outcome seems to have been a real shock to the event's American sponsors and reporters, though a close look would have told them that such head-to-heads had been organised before, with victory typically going to the *soroban*. *Stars and Stripes* called it a step backwards for 'the machine age'; the *Nippon Times* (controlled de facto by the occupying force) stated that 'civilization' had 'tottered'. Back in the USA, reports in *Time* magazine and elsewhere took a less affronted tone, but a note of surprise was still quite evident. Among writers on the *soroban* in Japan and elsewhere the contest remained a matter of comment well into the 1950s, and it remains part of the folklore of counting and calculation.

The fact is that a *soroban* expert can add and subtract numbers about as fast as they can be read out. For multiplication and division, informed comment in the 1950s reckoned that a first-class *soroban* worker was on a par with an electrical machine for problems with about ten or twelve digits: more and the machine would normally win, fewer and the human would.

When they were not being deployed in exams, competitions and public contests, the *suanpan* and *soroban* remained a common piece of equipment in shops and businesses throughout Japan, Korea and China. Compared with paper methods, they were quick; compared with the electrical machines that were now gaining attention, they were cheap and required neither regular maintenance nor an electricity supply. Their portability always remained a boon. Well into the 1980s, the bead-based devices could be called ubiquitous in East Asia. A degree of decline followed, as pocket calculators and electric cash registers displaced them as the most convenient tools for everyday calculation. But *suanpan*/*soroban* arithmetic remained an important part of mathematics education, and well into the new century a high proportion of adults reported using their methods at least some of the time. After-school training aimed at competitions retained its popularity too.

An intriguing feature of this culture was that the physical device itself proved to be, strictly speaking, unnecessary for the most skilled users. Regular users of the *suanpan* and *soroban* had long found they tended to interiorise the operations, to the point that they could do at least some calculations equally well on an imaginary device. By the second half of the twentieth century the so-called 'mental abacus' (*anzan*, blind calculation) had become a regular feature of training programmes, and it featured in the *soroban* examinations in Japan and, later, their equivalents in China. As long as the number of digits involved in the calculations remains manageable, it seems to be possible for almost anyone to acquire a rapidity in mental calculation by this means that would be astonishing in other contexts. An observer in the 1980s described a class at Dongyuan:

> Children go there after school and usually sit on long benches in large rooms filled with students. A typical mental abacus calculation exercise begins when the teacher, standing at the front of the room, raises his hand, whereupon the room falls silent in anticipation. The teacher then reads aloud a list of 20 three-digit numbers as fast as he can, so fast in fact that the numbers are almost unintelligible. The children

are silent, and the room is tense with concentration. After the last number is read, every hand in the room shoots up, and the teacher calls on one child to report the sum. Usually the child's answer is correct.

Because there is no need to move actual beads, the speeds attainable are even greater than with a physical device, and the development of 'mental abacus' training has allowed its users to dominate competitive events like the Mental Calculation World Cup. An expert can add ten ten-digit numbers in under fifteen seconds.

The phenomenon caught the attention of cognitive science researchers, and has been regularly studied since the 1980s. The mental abacus provides a way to count, and to represent and manipulate numbers, that is both entirely mental and independent of language, written symbols or even physical gestures. Experts can hold a conversation, or tap out a rhythmic beat, while calculating. There is evidence that proficiency with the *suanpan*, the *soroban*, or their mental equivalents, results in a person using different parts of the brain to perform numerical tasks compared with someone who has learned to calculate using words or symbols: motor centres rather than language ones. It is a helpful reminder that words and symbols are not essential to counting, even of the most sophisticated kind.

Herman Hollerith and Kawaguchi Ichitaro: Counting machines

What about the electrical machine that Private Wood operated, and that was so soundly beaten in 1946? Where does it fit into the story of counting, in East Asia or elsewhere? To find out, it is necessary to take a step back to the years around 1900, when modern anxieties were coming to dominate the very ancient practice of census taking.

Japan – probably Tokyo – in 1905. An operator sits at a prototype machine. Made of dark polished wood, it is the size of a large upright piano. The operator has a flat desk to work on, and faces a rack of forty dials, each one like a clock face, with two hands.

She inserts punched cards into the machine one after another, taking them from a dedicated card-sorter that stands to her right. She places a card in position, and turns a handle. The machine makes electrical connections through the holes in the card; the pattern of holes determines which circuits close and which do not. That in turn determines which of the forty dials advances a step: perhaps one, perhaps several. Each dial has markings up to one hundred: with two hands, each can count to ten thousand.

At the end of the working day – or the set of cards – the numbers can be read off and written down, and the dials reset to zero.

Kawaguchi's Electric Tabulation Machine.

The machine was designed by Kawaguchi Ichitaro, an engineer working in telegraph signalling technology, to assist with the tabulation of census data. The request for such a device came from the Japanese cabinet via the ministry of communications and transportation. His prototype was complete in 1905, and was used to process some of the data from the previous year's demographic survey.

The creation of this counting machine was one of the outcomes of a thirty-year agitation for more and better statistics about the Japanese population and economy, led by individuals from both

government and business. A system of land maps and family regis-
tration, very broadly similar to the old Chinese taxation surveys,
already existed, but after the Meiji restoration in 1868 the government increasingly perceived a need to gather more information
about the country's economic activity, to help plan its growth and
industrialisation. By the time a new national census law was prom-
ulgated in 1902, it had become clear that the volume of data to
be gathered would benefit from mechanical processing: hence the
request to Ichitaro to build his tabulating machine.

———

Those familiar with the history of computing and calculating
machines will have a sense of *déjà vu* about Ichitaro's machine. It
bore a marked resemblance, both in its appearance and in the way
it worked, to the machines designed in the USA over the previous
two decades by Herman Hollerith, and that were spreading rapidly
throughout the world during the decades around 1900.

In the USA, carrying out a regular population census had been
an obligation ever since 1787, when the Constitution specified that
the apportioning both of members of the House of Representatives,
and of direct taxation, should be based on the populations of the
various states in the Union. It specified that a census – an 'enumeration' of the population – should be taken for this purpose within
three years after the first meeting of Congress, and at least every
ten years thereafter. The first was duly carried out in 1790.

Over the following century, the details of the census and some
of its assumptions changed drastically. The population of the United
States grew from under four million to around sixty million, and
both government and people came to demand more and more
information about that population. From the initial handful of
common-sense questions about names and numbers, the work of
the census enumerators steadily grew in complexity, with new

questions added about agriculture, industry, births and deaths, schools, libraries, churches and property. The 1880 census asked twenty-four separate questions, produced hundreds of tons of raw forms ('schedules'), and resulted in the publication of twenty-two large volumes of data. Officials worried whether the situation was sustainable. A census is as much about reducing data to manageable form as about collecting it in the first place: any set of data can be processed given enough time, but there were limits to the delays that could be tolerated and the money that could be appropriated to fund the work.

'Counting' the census results, up to this time, had meant making tally marks on sheets of paper. Specifically, sheets were prepared covered with a grid of small squares, and in each square up to five pencil marks were made by a census employee who was counting, say, the number of census schedules that reported their subject as practising a particular occupation. A completed line of five squares contained twenty-five tallies; a complete block of four lines contained one hundred. Making the marks by inspecting the original census schedules in this way was repetitive, boring and susceptible to error, and converting the completed sheets of tallies into Arabic numerals, written on another sheet of paper, was also eminently liable to be done wrongly by tired, hurried or stressed human beings. Furthermore, if several kinds of data had to be extracted – age, occupation and marital status, say – the original census schedules had to be handled several times, creating a bottleneck in the whole process.

To avoid the expected disaster at the 1890 census, its director arranged a competition to test possible mechanical solutions to the problem of large-scale, repetitive counting. Mechanical aids had already been used over the previous twenty years, including a tally sheet that was automatically moved on rollers, and a system in which the raw records were initially transcribed onto more easily handled separate cards or papers. At the trials in 1889, systems were shown involving paper slips and coloured inks, or cardboard 'chips' of different colours. But the competition was easily won by a system using punched cards to transcribe the census data, invented by a 29-year-old son of German immigrants, the Columbia graduate Herman Hollerith.

Punched cards or punched paper strips had long been used to control mechanical looms and player pianos, or to store information about the holders of railway tickets. Hollerith's innovation, prompted at least in part by a suggestion from his fellow census employee John Shaw Billings, was to use a punched card to store information about each individual recorded in the census. Crucially, he realised that information in that form could be sorted and counted automatically by a mechanical or electrical device. By the time of the 1889 trials for the federal census, his system had already been patented and successfully used to tabulate health statistics for Baltimore, New York and New Jersey, and for the US Army.

Starting from Monday 2 June 1890, nearly 47,000 census enumerators went door-to-door across the USA, gathering data. There were eight million square kilometres to cover. One printed form was used for each family, with twenty-six questions to answer about each person, and five extra questions about the household. Additional forms were used for veterans, for agriculture, manufacturers, the disabled, homeless, convicts . . . a total of nearly seven hundred different pieces of information were collected. There were penalties for refusing to answer. Special agents, meanwhile, were separately collecting information about property, mortgages and a huge range of industries and products: from agricultural implements to hosiery, from brick yards to silk goods.

Each enumerator sent daily reports of work completed to a supervisor, as well as sending in the completed forms themselves.

After checking, the forms went to Washington:

> The blanks which had been filled up were laid one upon the other on a piece of straw board. Each pile contained the schedules of a single enumerator. On top of all was placed an empty portfolio, to whose center was pasted the label with the enumerator's name and the designation of his district upon it. The bundle was then corded together and a number of such bundles, representing from 13 to 15 enumeration districts, were placed together in a box which they exactly fitted. The box, 27 inches long and about 18 inches in its other dimensions, properly closed and sealed, was sent in this shape to the Washington office. One hundred such boxes were received daily, and several trucks were kept busy transferring them.

At the Washington census offices, a total of something like twenty million forms were eventually received, weighing around two hundred tons. The data on the forms was transferred to punched cards, using a device much like a typewriter keyboard rigged to a hole punch. One card was prepared for each person surveyed: thus about sixty million cards, each about the size of a dollar bill, with an average of perhaps twenty holes per card recording different pieces of information according to a precisely defined scheme of hole positions. With practice, a person could punch a thousand cards a day, and it was reported that by the end of the process the clerks had handled so many cards they could 'read' the information from them just by looking at the holes.

After six months the cards were ready to be counted. A journalist reported visiting 'a very tidy and airy machineshop . . . where nice-looking girls in cool white dresses are at work at the long rows of counting machines' that 'remind one of upright pianos'. One of these clerks would place a card flat on a hard rubber plate and pull a handle to bring down a frame on top of it. A grid pattern of electrical points in the frame met a similar grid of mercury contacts in the plate, making contact in just those positions where there was a hole in the card. Circuits were thus completed, a bell rang, and a selection of the machine's forty dials would each advance by one. A quick-fingered clerk could process 7,000 cards in a day, and there were up to a hundred clerks.

A Hollerith tabulator in use in 1890.

A preliminary count of the whole population was completed by 12 December, just over a month after the last census returns came in. A total of 62,629,250 people had been counted. Subsequent work provided detailed breakdowns of this number by state, city and town, and similarly broke down the count of people by their place of birth, age, occupation, parentage, literacy and marital status. The electrical circuits could count a complicated pattern of holes just as easily as a simple one, and they could count several such patterns simultaneously: up to forty in principle, as many as there were dials on each Hollerith machine. Furthermore, when the circuits closed they not only rang a bell and advanced certain counters; they also performed the final task of opening the lid of one of two dozen boxes to the side of the machine. The clerk took the card out of the machine, put it in the open box and closed the lid. Thus, as one set of counts progressed, the cards were automatically being sorted ready for the next set of counts.

Journalists who visited the workroom were invariably impressed, if half deafened by the bells ('. . . your tympanums all tingle / At the jingle, jangle, jingle / Of the Bells!'). Two dozen volumes of statistics were printed over a period of seven years, and the census of 1890 was unarguably the largest and most detailed view of the American population and economy yet produced (notwithstanding a widespread belief that certain cities, and perhaps the population as a whole, had been under-counted). It was one of the printed bulletins from this census that famously noted that 'the unsettled area' of the American continent 'has been so broken into by isolated bodies of settlement that there can hardly be said to be a frontier line', prompting a long-running conversation about the meaning of the frontier – its presence, its movement and now its absence – for American history. The census also resulted in an increase in the size of the House of Representatives from 325 to 356 members.

Hollerith's machine was a genuine mechanical counter – not a machine for doing arithmetic, but one whose sole purpose and capability was to answer the question: *how many?* No human being had to pay individual attention to the items counted; they could simply be fed into the machine and a count accessed some time later, in the form of a number symbol indicated by the machine. The heart of the counting process – repeated attention plus keeping track – was carried out by the machine itself, raising an interesting question about how or whether a machine can 'pay attention' to anything. For the purpose of the census, the items counted were people as represented by punched pieces of manila card, but it was clear that almost anything could be counted by similar means. This was a genuine novelty in the history of counting, and one that, naturally, proved capable of enormous development over the subsequent decades. It is, perhaps, no great surprise that counting machines went on to change the world.

Machines of substantially the original design continued in use for processing the next US census, but from 1910 the Census Bureau ran its own machine shop to manufacture similar devices rather than rent from Hollerith at $1,000 per year for each machine. Hollerith's designs, meanwhile, were taken up for census work and other statistical tasks in a succession of countries including Austria, Canada, Norway, Russia, France and Italy. By 1912, an admirer could speak, admittedly with some exaggeration, of their adoption by 'nearly every civilized country'.

Later improvements included a system to record decimal numbers on the cards – using a choice of ten holes for each digit – and machinery to add up numbers so stored. Subtraction, multiplication and the means to represent negative numbers followed in due course, as did a device to feed cards into the machine automatically, doing away with much of the need for dextrous human operators. Special control cards could tell the machines to stop, and printing devices could automatically print the totals so that humans did not need to transcribe them.

Hollerith sold his limited company in 1911, staying on as a consultant; it became part of the Computing-Tabulating-Recording Company, which in the following decade changed its name to the International Business Machines Corporation. For much of the twentieth century, IBM would be one of the world's largest data-processing companies, supplying equipment to governments and businesses around the world.

———

There is a disastrous epilogue to the 1890 census itself. In a departure from the practice of the US census up to 1870, no copies were taken of the census returns before they were sent to Washington. One reason was to save labour, and to reduce the burden on local administrations of producing and storing literally tons of paper forms. Another was the belief that the punched cards themselves constituted an adequate copy of the data: one official wrote optimistically that the original schedules 'might every one of them be burned up, and the Eleventh Census could be taken over again from beginning to end, by means of the little slips of manilla': that is, the punched cards. In fact, that was not the case, since the names of individuals were not recorded on the cards, only reference numbers denoting the original schedules where they could be found: data was irreversibly lost during the transcription onto punched cards. A third reason was that some felt the original data itself no longer had value once it had been counted, reduced to aggregate statistics and those statistics published: an ill-omened side effect of too-efficient counting practices, perhaps.

By March 1896, an accidental fire had already destroyed portions of the original 1890 census schedules. The bulky remainder were still in the portfolios in which they had been submitted, tied up with twine and piled flat in racks. Limited funds were available for their preservation, and together with 180 tons of punched cards, they were stored in poor conditions in a basement in Washington. On 10 January 1921, a five-alarm fire tore through the building. There were no human casualties, but four hours of fire and the jets from twenty hoses damaged perhaps two-thirds of the 1890 records beyond repair.

The remainder were disposed of as useless in the 1930s, missing out very narrowly on the advent of new attitudes to records management and the preservation of historical documents. The cornerstone of the American National Archives building was laid just a day before the disposal was authorised, and a programme to photostat and bind old census records was under way within the year. It was too late for the 1890 schedules and cards, and with their destruction a permanent, much-lamented gap was created in the records of American genealogy and social history.

What of the Japanese census machine, based on or inspired by Hollerith's long-lived invention? By February 1904, Japan was at war with Russia, in a conflict that continued for a year and a half, with huge consequences for the government's priorities and budgetary allocations. The census was postponed indefinitely, and a national population survey of Japan was not taken again until 1920, by which time Ichitaro's prototype machine was obsolete, perhaps forgotten.

Today it rests in Tokyo, in the Statistical Museum of the Statistics Bureau of Japan, alongside other early computers and tabulators. Who were the people this machine counted, during its brief period of use as a prototype? It seems impossible to say.

Sia Yoon: Counting likes

South Korea; spring 2022. Sia Yoon (five feet tall, blood type A, born on 15 March and now seventeen years old) is woken by her digital alarm clock, sits up in bed, stretches. Turns on music (Schubert), showers, dresses, leaves her tiny apartment for Dara High School.

Half a million people follow Sia's morning routine, because she is a successful vlogger. Aged seven, she was on TV as a martial arts prodigy, and she has never looked back. She calls herself a 'natural-born cutie' ('Yes, yes! Be jealous of me!'), and the further reaches of stardom now beckon. A million followers is a goal, and she is starting to produce commercially sponsored posts: for lip tint, for tote bags (the tote bag post got 24,318 likes).

Sia's day continues with classes, lunch, meetings with friends. She posts photos, checks the responses: how many views, how many likes (or dislikes). Checks her ranking ('Of all Dara High School students, I rank in the top 1% for the number of subscribers!')

It's not all good news. Sia recently broke up with her boyfriend and vlogging partner of two years. There was naturally an element of personal despair about her response, but she's also deeply worried about the effect on her online presence. Her photo post announcing the breakup got 155,205 likes in a few hours, but her subscribers fell away in their tens of thousands over the subsequent days. Her life at the moment is dominated by the search for ways to remedy the situation.

Sia Yoon, star of NewTube.

In the decades either side of 2000, culture was transformed in many countries by a set of processes loosely grouped as 'digital'. The hardware involved descended from three sources.

On the one hand, there were mechanical calculating machines, whose history goes back to the seventeenth century if not before. By the close of the nineteenth century, desk calculators that could perform the four arithmetical operations were commonplace, and they remained so until the middle years of the twentieth. Second, there were mechanical counting and calculating devices like Hollerith's, using punched cards to store data, and manipulating the cards to perform various operations upon that data. Third, the typewriter and – even more common at one time – the cash register: input devices more user-friendly than laboriously punching holes in cards.

In the 1930s, it was already becoming feasible to assemble multiple punched-card machines to perform and control complex sequences of operations automatically, for applications in fields such as astronomy and engineering. IBM marketed a 'card programmed calculator' in the 1940s and 1950s that was already effectively a computer in the sense of being programmable and able to store, retrieve and process data. Innovations in the basic hardware followed – vacuum tubes, transistors, integrated circuits – leading over a few decades to smaller, faster and cheaper machines. Punched cards, as a means to store data and control machines, were similarly replaced by magnetic tapes and floppy discs. But the basic concepts of data storage, data manipulation and program remained central to what a computer was.

A vacuum-tube computer in the 1940s and 1950s filled a room or even a building. Their names – they were rare enough to have individual names – now sound like quaint jokes: 'Colossus', 'Whirlwind'. Yet, in the mid-1950s, one of them – the ENIAC at the University of Pennsylvania – could plausibly claim to have done more arithmetic in its eleven-year life than humanity had done in the whole of its previous history.

The uses of computers at first remained limited to industrial, military and scientific calculation, and to tasks of the bookkeeping type: airline reservation systems, human-resources administration and stock control. In the 1960s, there were perhaps 10,000 computers in the world. The following two decades saw the rise of affordable computers for individuals, and a software market sprang up, focused in particular on word processing, spreadsheets and databases. User interfaces became graphical rather than text-based in the late 1980s. Meanwhile, microchips found an increasing range of uses outside computers: in cash machines, for instance, in barcode scanners, in digital watches, in arcade games.

Initially, computers were isolated devices, but that, too, gradually changed through the 1960s and 1970s as new capabilities were devised. First, to enable remote working from multiple terminals within a single university campus, all linked to one central computer. Next, to network together multiple computers, whether in local networks with dedicated wiring or – using existing telephone lines – across continents and beyond. Once protocols had been standardised for securing a network and exchanging data over it, commerical online services took off rapidly from the mid-1980s. In 1990, there were just over 300,000 computers on the internet; a decade later there were 100 million.

Email was the crucial application in the initial stages of the internet's popularisation, closely followed by the World Wide Web and by social media platforms which hosted, aggregated, sorted and sifted ever-increasing oceans of user-generated content. By the 2020s it was becoming difficult to think of a social or economic function that could not be done online: shopping, meeting, gaming, banking . . . The growth of the internet coincided with the development of more and more portable devices – laptops, smartphones, tablets and smart watches – and drove the spiralling demand for them.

Since the 1930s, science-fiction authors had speculated and fantasised about the global sharing of information, under evocative titles such as the 'world brain', the 'memex' and the 'noosphere'. In the first decade of the twenty-first century, it seemed rather as though those dreams had come true, with both their positive and their negative consequences. For some, the internet created a radically new form of culture, a 'consensual hallucination', a digital play-ground in which individual perception, participation and bricolage would replace the old modes of cultural production. An ever-wider range of cultural artefacts were now either being 'born digital' or were rapidly being converted into digital surrogates for online distribution. Online identities were freely malleable, online free-doms unconstrained.

At the same time, it was conspicuous that some parts of human experience were being left behind because they resisted being trans-formed into data, and that some people – indeed, most people – were being left out of the digital revolution because of the acci-dents of wealth and access to technology. The new, participatory online culture was seldom – if ever – as open, as unconstrained or as democratic as it at first seemed, and it facilitated negative behaviours – theft, lying and bullying – just as often as positive ones. It also facilitated spying, coercion and control, sometimes in new and alarmingly powerful forms. In the new realm of online data, it often felt as though the tasks of navigation and quality control had been outsourced to end users ill-equipped to perform them: that what had been created was an unstable, unreliable post-truth society in which what mattered was how many followers you could get to believe you, not whether what you said bore any relationship to an offline reality.

All of this has everything and nothing to do with the history of counting. What is digitisation if it is not, in a sense, 'counting'? (The use of 'digit' – literally a finger – to mean a number or a number symbol seems to derive from Roman and medieval finger-counting.) What is online culture if it is not the sheer profusion of modern ways of counting, the participatory spectacle of a world turned into pure numbers? Computers were inherently digital from the days of vacuum tubes, consisting of electrical components whose state was either on or off, conventionally interpreted as 1 and 0 in a binary representation of numbers.

Yet even by the early years of the twentieth century, punched-card machines were performing tasks very much more complicated than just counting. It seems forced to describe, say, the mid-century code-breaking devices as mere 'counting machines', and frankly bizarre to insist that that is what a smartphone really is. Indeed, once the function of counting has become detached from having a human being pay sequential attention to things or events and keep track while doing so, machine counting quickly begins to feel very unlike other kinds of counting. The interesting questions shift to its capability for transforming people into punched cards, paintings into on-screen images, videos into advertising revenue: and away from the fact that those transformations are mediated by binary numbers.

That said, microchip devices are also used for functions much more like old-fashioned counting: as a substitute, when asking *how many*, for counting in words, on your fingers, or using some other device. One of the most conspicuous is the counting of friends, likes, followers and views in social media contexts. Here, the digital world has provided not only a new technology with which to count, but new things to count.

———

All of these issues play out in microcosm in the life of Sia Yoon. She lives in South Korea, a country whose industrialisation and economic growth have involved conspicuous success in the electronics industry, from semiconductor components to phones and watches, and also successive phases of cultural export – the 'Korean Wave' – facilitated and driven by online marketing and online presence.

K-pop leads the way, with the most popular artists and videos gaining countless followers online. By December 2014, Psy's video 'Gangnam Style' was on track to 'break YouTube', as the number of views approached the largest that can be stored using thirty-two binary digits (rather more than 2.1 billion, assuming you use the first digit to record whether the number is positive or negative), neatly illustrating that every way of counting has its limits. YouTube upgraded its hit-counter to use sixty-four digits instead, and the problem was averted. Games, films, television and of course devices by Samsung and LG are also major exports. Korea is one of the most wired countries in the world in terms of broadband and wireless connectivity and ownership of digital phones, cameras and other devices.

After her breakup, Sia found a new boyfriend fairly quickly, and started a vlogging project that involved making him over. Minho (five foot ten, blood type AB, birthday 17 August) was, to begin with, a nerdy devotee of Japanese animé: overweight, unhealthy and uncharismatic. His best friend was shocked at the very idea of liking a girl who lived 'outside the monitor'. But Sia was convinced that this rediscovered childhood crush could be transformed: specifically over the course of one hundred days, which she diligently counted down on camera as her followers watched. The project was a winner on social media: her initial announcement was viewed 1.59 million times in the first day, receiving 27,000 likes (and few or no dislikes), and she went on to vlog in detail about Minho's newly imposed exercise and diet regime ('W-wait . . . But . . . There's nothing but leaves and sweet potatoes here . . .').

His first workout video got 250,000 views in a day; a weigh-in 880,000, with 3,000 comments. As his appearance improved, the couple's first selfie together gathered 20,000 likes, and a clip from their skiing trip went viral and attracted 1.6 million views. And so on, and so on. The project succeeded in terms both of Minho's health and appearance (his weight eventually dropped to 169 pounds), and of subscriptions, which rose to over 650,000. As one supporter put it, 'you've become an icon of hard work and effort, and everyone loves you for it'. Yet Sia would never lose sight of the fact that 'what's important is the view count'.

—·—

Sia, perhaps fortunately, is not a real person. She is a character in the webtoon *My Dud-to-Stud Boyfriend* (story by Yerang; art by Sutggi; nine million page views at the time of writing).

Webtoons descend from printed manga and from the online picture-diaries of the 1990s. Like K-pop, they have surged hugely in popularity over the first decades of the new century, both inside and outside Korea: by 2018, over 35 billion episodes had been read, and half a million views per episode was not uncommon for a successful series. Commissioned and volunteer translators render them into – so far – over thirty foreign languages. Films, TV shows and games have appeared as spin-offs. As scholar Dal Yong Jin describes:

> on a bustling subway ride to work, it's not hard to find people staring into their smartphones or PC tablets, scrolling quickly down the screen to see what comes next. On sidewalks or in coffee shops, the situation is the same. If a smile can be spotted on the user's face, it would not be far-fetched to say they could be reading one of their favorite webtoons.

Adapted to presentation on the web or through a dedicated app, webtoons scroll vertically through sixty or eighty panels per episode, often ending on a cliffhanger. By contrast with printed manga, they are presented in colour, sometimes with the addition of sound or animation. Genres range from biography and sports to crime noir and fantasy. Romances and self-consciously mundane slice-of-life stories are particularly popular, in which self-referential appearances of technology and social media are common: characters use phones to count blog posts, likes and comments, smart watches to check heart rate and time exercises, tablets to access music, video and news and shop, check the time, track bank payments, read toons . . .

The so-called artist incubation system, in which artists compete for audience attention on the webtoon platforms, has much in common with social media platforms like YouTube (or the fictional equivalent NewTube, on which Sia Yoon is such a star), with comments, likes, hit counts and followers. There has been both criticism and celebration of a model that lowers the barriers to entry – 'webtoon artist' is reportedly one of the most popular career ambitions for Korean pre-teens – but concentrates the rewards on a very small percentage of participants, picking a handful of winners based mainly on their popularity, who are then under intense pressure to conform to audience expectations and keep up a gruelling schedule of new material. Just like Sia herself, in fact.

Sia's story is perfectly believable in a digital world, one in which what counts is what you can count, or rather what your many devices can count for you. In which success, popularity and worth are matters of automatically aggregated numbers of hits, views and likes. The resonance between her story and the 'attention economy' in which webtoon artists themselves operate adds a layer of richness to her; and perhaps a whisper of unease too.

Sia's story is not yet finished, and her many fans – both real and fictional – eagerly await what will happen. Now that her new boyfriend is transformed, she and he have taken up positions of fame and social leadership in their school, but they have also rapidly learned of the costs of exploiting social media to shame and expose their enemies. The sponsored posts continue, but so do the dilemmas; so, at times, does a certain emptiness at the heart of Sia's digital life. Her birthday party photos get 10,000 likes, but she comments ruefully that 'all the presents I get every year are just sponsored products from brands'. According to Sia's creators, there are years of content still to come, and there is no doubt that she will go on compulsively, mechanically counting views, likes and followers, all tracked for her by her devices and the online platforms to which she subscribes. She is, in more senses than one, the product of a digital age.

———

Is a digital world one that contains more counting, or less? What does the future hold, as human beings outsource more and more of their everyday counting (and arithmetic) to sophisticated machines? Is it reasonable to imagine that arithmetic, and even counting itself, might eventually vanish from human practice, in the same way that, say, copperplate handwriting has done? Perhaps it is. Few people with access to digital technology would now do more than the simplest arithmetic either mentally or on paper, rather than asking a device. Few would count more than, say, a few dozen objects by hand without outsourcing the task to a machine. When was the last time you counted more than a hundred of anything? More than ten?

Perhaps one of the possible futures of humanity is a world *after counting*.

7

Counting words and more in the Pacific world

Counting machines are another branch of the counting story that is in danger of feeling like *the* story; their present-day forms are so dominant as to overshadow any other way of counting, past or present. But they are in reality only one branch among many. To continue the journey around the globe is to find further worlds of counting, in the Pacific and the Americas: rich, distinctive and fascinating.

———

After walking as far east and south as mainland Asia will take you, keep going. Build rafts, build canoes, and paddle or sail across the channels of water to island after island. Through the archipelago of Indonesia, making short sea passages on which you can see the next island from the one you just left. On a few of the longer passages, your destination is out of sight: a new sort of voyage into the unknown.

Continue as far as the wind and the water will take you. Discover forests, deserts, barren atolls and teeming reefs. Find new worlds, and devise new ways to live in them. To the great continent in the south; to the scatter of tiny islands beyond. Carry your language, your culture – your ways of counting – with you.

From start to finish, it would take human beings over 60,000 years to populate the Pacific.

———

Not surprisingly, Oceania is home to an enormous diversity of human cultures, and therefore of ways of counting. The extraordinarily deep human history in the Pacific world, with its successive waves of migration over tens of thousands of years, give the story of counting a particular flavour here. These were cultures that relied on spoken, not written words: their history must rely on archaeology and on the memory of living languages and communities. There were no number symbols here, but instead an extraordinary range of number words and gestures, from some of the simplest systems documented to some of the most complex. That range is well represented by languages spoken today. The languages of Australia, spoken by descendants of the first humans to arrive in the Pacific, have on the whole quite simple counting systems, sometimes consisting of just a few words. The peoples of neighbouring New Guinea, on the other hand, are famed for elaborate routines combining word and gesture, which seem to have developed *in situ* over perhaps the last few thousand years. Finally the Austronesian languages, whose speakers populated Micronesia and Polynesia over the last 5,000 years, use regular decimal systems to count to extremely high numbers, used for food distribution and other purposes. This is a region that can display the range and power of human counting words like no other.

Ayangkidarrba: Counting eggs

Ayangkidarrba (Groote Eylandt), early in the twentieth century. A woman has gathered turtle eggs, and she returns to camp in the early evening, carrying them in a paper-bark basket. Others have hunted, fished, gathered. She shares out the heap, counting in fives: *awilyaba, ambilyuma, abiyakarbiya, abiyarbuwa, amangbala . . .*

Groote Eylandt lies about 50 kilometres off the Australian coast, at the western side of the vast Gulf of Carpentaria. It has been an island for 7,000 years or more, and the original inhabitants of its 2,000-odd square kilometres are the Anindilyakwa. Hills, springs, rivers and of course the coast make for a rich range of habitats from forest to sand dunes. Traditionally, it supported a few hundred people, at one of the highest population densities in aboriginal Australia.

The island was a dense network of named places: natural features and resources controlled by different clans. The human world was a similarly dense network of kinship relationships linking fourteen clans. A prodigious knowledge of the local environment enabled the Anindilyakwa to gather foods ranging from berries, roots and fruit, wild honey and drifting coconuts to wallabies, dugong, fish and turtles, in a pattern changing with the seasons. Spears, harpoons,

Sea turtle eggs, as counted on Groote Eylandt.

bags and baskets, ornaments, musical instruments and even canoes were all made from local materials including wood and bark.

An ethnographer visiting in the 1970s described a visit to the bush, by then much rarer than in earlier generations:

> [the people] would awake at sunrise, boil some tea and have a bit to eat from what had been left over from the previous day. Then they would work fishing, hunting or gathering until the sun was directly overhead, eating enough to sustain themselves from what they could gather as they went along. During the heat of the day they would sleep, then begin work once again in mid-afternoon, returning to camp in the early evening with the bulk of the day's catch, which they would then prepare and consume communally. Following this they would sometimes wander through the shallows with a light in search of fish, or rest beside their campfires four abreast with a fire between each pair telling stories. After a few hours of sleep they would again wake up, have something more to eat, wander around and perhaps tell another story, then sleep again, waking at sunrise.

The Anindilyakwa had strong relationships with their neighbours on the smaller islands surrounding Groote Eylandt, and on the mainland; they imported certain materials – flint, ochre, emu feathers, bamboo – in exchange for local products, and they inter-married. And for two hundred years or so, up to the early twentieth century, they were regularly visited each December by Makassar people from Sulawesi in Indonesia. Ranging widely on the north Australian coasts, these visitors harvested sea cucumber (boiled, dried and smoked, it made food and medicine prized in south-east Asia); they also took turtle shell and pearl shell. Employing the Australian people to work for them collecting sea cucumber, they provided such imports as tamarind and chilli, cloth and metal items. They introduced the dugout canoe, some new place names, and perhaps forty words, mostly for the items they themselves brought.

The Anindilyakwa's own language – Amamalya Ayakwa – spoken on Groote and the smaller islands nearby, is probably related to languages spoken on the nearby coast, but long development in separation from the mainland has given it a highly distinctive character. Words are long – up to fourteen syllables – and there are thirty-two distinct consonants. The grammar is one of the most complex ever documented, with nouns divided into eight classes which require different prefixes on the corresponding verbs and adjectives. As well as singular and plural options there are also special forms of verbs, nouns, pronouns and adjectives for the dual (two items) and the trial (three items, sometimes four).

Meanwhile, the actual counting words work in fives, with words for 'one', 'two', 'three', 'four', 'five', 'ten', 'fifteen' and 'twenty'. They gain prefixes to agree with the nouns they accompany, like other adjectives in the language. There are special forms for 'once', 'twice', 'three times' . . . and for 'by ones' 'by twos', 'by threes' . . . Sometimes gestures support the counting routine:

The hand is held loosely with the palm facing the person counting. The fingers are placed together one by one; index finger to thumb, middle finger to thumb, ring finger to thumb and little finger to thumb, until all fingers are bunched together. If the number is more than five the fingers are held together while counting continues with the other hand. After ten, the toes are touched one by one, first on one foot and then on the other.

Despite this, the number words show no detectable relationship with the terms for hands or feet.

What were the counting words used for? One context was hunting: counting numbers of people available to hunt or numbers of animals brought in. Similarly, the number of warriors going out to fight or returning might be counted. Another possibility was calendar calculations; yet another, the counting of food resources like eggs. In fact, the main context for counting in Amamalya Ayakwa seems to have been sharing mid-sized food items like fish, eggs or wild apples.

There is a story on Groote Eylandt about a mythical dog that could count. The story tells how the dog went hunting and returned to his family with the turtle eggs he had found. He began counting them in order to share them (though the story reveals that the dog cheated by hiding some eggs, causing a fight to develop!).

Reports from the mid-twentieth century show that turtle eggs could be gathered in quantities of over a hundred, and the counting procedure, as demonstrated to a visitor using a heap of pebbles, showed how they were counted out into heaps using the base-5 number words.

The Anindilyakwa were both typical and atypical, on a continent where change and diversity seem to have been the norm. Australia was populated – following a rapid human expansion through Indonesia – at least 55,000 years ago, perhaps more than 65,000. At that time, sea levels were much lower; large parts of island Southeast Asia were connected to the mainland as the single land-mass of Sunda, while Tasmania, Australia, the Torres Straits and New Guinea were joined as Sahul. Nevertheless, some significant stretches of open ocean had to be crossed to reach Sahul: passages of 20 kilometres or more. Genetic evidence shows that there was a small degree of interbreeding with the hominids already present in Asia and Indonesia. It also shows that the peopling of Australia was essentially a single event, with no substantial new incursion into the aboriginal population after the initial period of settlement. The Australian cultures and languages were thereafter substantially isolated for tens of thousands of years.

That is an enormous depth of time: as in the African Stone Age, it is more than long enough for words, clusters of words, technologies and practices to have been invented, forgotten and invented again, perhaps many times. In an oral culture, it need take no more than a few generations for a word or an idea to be forgotten, leaving no trace that it ever existed. Of course, a degree of cultural stability is to be expected, but there is no sense in which Australia was either a static environment or an unchanging set of cultures. Shifting coastlines and gradually changing climates combined with the creativity and restlessness of human beings to make the conti-nent a shifting mosaic of different lifeways, from the tropical north to the arid central deserts, from the temperate southern coasts to the sinking islands and land bridges of the Torres Strait.

A reliance on hunting and gathering united the continent, as did the use of a broad range of animal and plant species, from fruits, seeds and wild honey to wallabies, fish and geese. Tools ranged from harpoons and axes, fishing lines and lures, to spoons, bags and cloaks. Fire was used for warmth and cooking and to manage vegetation and species diversity, with some plant species deliberately planted and reseeded.

A tradition of painting on rock likewise spanned the continent, with a range of regional styles; beads of shell and bone were worn, and each tribe had its stories handed down across unknowable gulfs of time. And Australia was also crossed by networks of trade, each tribe interacting directly with its neighbours and indirectly with the sources and destinations of goods, stories, ideas across long distances, even as far as the Torres Strait Islands and Papua after they split from the mainland.

———

By the eighteenth century, there were perhaps five hundred tribes linked in the Australian network of trade and lifeways. They spoke 240 or 250 distinct languages; many people spoke two or more. It is likely, though not certain, that nearly all of those languages descend from a single ancestor: but the depth of time for which they have been changing is huge and has facilitated much mutual borrowing of words, sounds and structures as well as much evolutionary change. No languages anywhere else in the world can be shown to be related to the Australian languages.

Like any languages anywhere in the world, each of the Australian languages stands at the end of a development of tens of thousands of years whose roots lie in Stone Age Africa. Like their cultures, they are not in any sense 'primitive' languages; their vocabularies are as large as others and their grammars are if anything more complex than most. They tend to have many distinct consonants, but few – typically three – distinct vowels. Large or complex systems of noun classes are common, as are certain phenomena of noun case which are rare elsewhere in the world.

It is important to know that Australia's cultures and languages are not 'primitive' – not 'relics' – in order to approach their counting systems, which are also distinctive compared with those of the rest of the world. When the languages were documented in the nineteenth and twentieth centuries, three-quarters of them possessed just three or four number words. Five per cent of them counted only as high as 2. For the remaining 20 per cent, counting limits from 5 to 100 were reported. Observer bias may have played a role in underestimating the extent of Australian counting, as may the fact that many of the languages were already badly disrupted by the time their words were written down. But still, compared with other parts of the world, Australian languages show a tendency to low limits in their sequence of counting words.

The internal structures of the lists of counting words show a striking diversity from language to language. While a proportion of the Australian languages show no evidence of any base system – in one case, a set of nineteen counting words appears to have no internal structure at all – the majority of them do extend their verbal counting routine by combining smaller counting words into larger ones. But decimal systems are rare here. A list of two, three or four numbers cannot possibly have a base-10 structure: but even those Australian languages with larger sets of counting words are remarkable for the absence of decimal systems. Most in fact use 2 as a base, but systems with bases 3, 4 or – particularly – 5 are also found.

In a final twist, some of the languages with very small sets of counting words seem sometimes to have extended them using gestures, counters or tallies: expressing, recording or calculating with numbers for which the language had no specific word. Reports from around Australia show that certain kinds of counting tasks tended to be performed using fingers in routines like the one on Groote Eylandt. In languages whose number words stopped before five, displaying a hand could perform the functions of the fifth counting number.

The Anindilyakwa were part of a complex mosaic of different ways of doing things: small number systems, large number systems, systems with or without a base structure, systems extended or not extended by the use of gestures. But even this does not exhaust the possibilities to be found in Oceania; for elsewhere in the Pacific world, counting by gestures attained even greater importance and sophistication.

Oksapmin: Body count

The central highlands of Papua New Guinea, 2012. A man walks into a small store and buys some food items using the local currency, *kina* and *toea*. Prices are written up on a board behind the counter.

He and the proprietor use coins and number words to negotiate about what must be paid and what change must be given. They also use a complex, rapid series of gestures, pointing to a conventional sequence of body parts around the hands, arms and head; the gestures are just as important as the words, perhaps more so. In fact, the words are simply the names of the body parts.

To add two numbers together or subtract one from another, it is necessary to repurpose these gestures in ways which the participants' ancestors would not have recognised. But both men leave the transaction satisfied that the correct number of coins has been exchanged.

North of Australia, south of the equator, New Guinea is the world's second largest island, home to a multitude of species, including human beings. It was populated by humans many thousands of years ago while it was still connected to Australia, but the central highlands remain remote to this day, with arduous trails leading in by foot and a few scattered airstrips for single-engine planes. A rugged geography of mountains, rivers and gorges, as well as internal warfare, have contributed to the isolation of the human groups living here.

One of those groups is the Oksapmin. Their territory in the mountains near the centre of the island is bordered to the west by the Victor Emanuel mountain range (cold enough to kill), to the north by the Ok Om river (a raft crossing described as 'very risky') and to the south and east by the Strickland Gorge, sometimes compared to the Grand Canyon and involving a cliff of nearly 500 metres which must be scaled with the help of wooden hand-and-foot holds. It is a beautiful landscape, with sparkling valleys, sandy rivers and smoke creeping from thatched roofs. A chequer of gardens fills the land; clouds garland the forested peaks. The climate is tropical, with wet and dry seasons, and temperatures characteristic of the highlands, down to ten degrees during the colder nights, up to the thirties on the warmest days.

In the traditional Oksapmin lifestyle, the days are filled with work in gardens of taro and sweet potato – repairing fences, clearing weeds, harvesting – punctuated with tending pig and hunting in the forests for birds and small animals. Tools are of wood, bone and flint, supplemented over the last half-century by imported steel knives and axes. Houses are of pandanus bark, roofed with leaves, and they scatter the valleys in groups of two or three. The total population of Oksapmin speakers was estimated in the 1960s at around 4,000 people; today, it may be more than twice that.

Trade was important to the prehistoric life of the Oksapmin and their neighbours: they obtained stone axes, shells, bows, salt and drums from their neighbours, and shell valuables were used as a kind of currency across groups in the mountains. Goods could travel hundreds of kilometres, despite the boundaries of geography and language.

The Oksapmin counting sequence consists of gestures towards twenty-seven parts of the body. It begins at 'one' on the thumb of one hand, and moves across that hand to the little finger ('five'), with the index finger of the other hand being used to point. Next, it continues up the arm – wrist, forearm, elbow, upper arm, shoulder, neck, ear, eye – finally reaching the nose ('fourteen'), the midpoint of the count. The count then continues symmetrically down to the other wrist, with a switch in the hand being used to point. The final five elements – 'twenty-three' to 'twenty-seven' – run from the thumb of the second hand to its little finger, meaning that the end of the sequence is not quite the mirror image of its beginning.

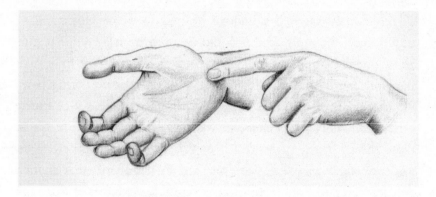

Showing the number 6 in Oksapmin.

Each item in this sequence also has a spoken word – *tipana*, *tipnarip*, *bumrip*, *hadrip*, *hatatah* – which is identical with the name of the body part; context makes it clear whether the body part or the counting number is intended on any given occasion. A prefix (*tan*) is added to the names of parts on the second side of the body to distinguish them from those on the first. Some reports indicate that the words can be used on their own to count, but most descriptions of the system regard the gestures as essential.

The complete cycle ends at twenty-seven – second little finger, *tan-hatatah* – and may be denoted by the word *fu*. If it is necessary to count slightly further, the system may be extended: not by starting again at the beginning, but by looping back from the little finger to the wrist of the same hand ('twenty-eight'), and proceeding back up the arm. To interpret the resulting – potentially ambiguous – words and gestures is again a matter of context.

A separate set of words for the numbers from one to five also exists: the so-called conversational counting words, which correspond to nothing in the body tally. They run *tit*, *yota*, *yetir*, *yota yota*, *hanen*; they can be combined with the body tally if it is necessary to count more than one complete cycle: *tit fu* (one cycle), *yot fu* (two cycles), and so on.

The system was well suited to the uses to which it was put in traditional Oksapmin life. It could be used in any everyday situation where the number of elements in a set needed to be indicated. It could count valuables like shells or pigs, measure string bags, indicate a set of positions along a path. In daily life, a counting system of twenty-seven elements was more than adequate to count these things in the quantities in which they existed. There is no evidence that the system was ever used in arithmetic, that anyone ever learned addition or multiplication tables in terms of the body parts or their names. Indeed, there is no evidence that in traditional Oksapmin life any arithmetical computation was done at all.

———

Despite containing fewer than ten million people, New Guinea is home to around 850 languages, a sixth of the world's total. It possesses one of the most dense and complex linguistic situations in the world, the outcome of migration and mixing over tens of thousands of years, beginning with the first population of New Guinea and Australia. The mountainous region that includes the

Oksapmin area is home to several different language groups; Oksapmin itself is usually considered an isolate, with no close relatives among living languages.

The region also has probably the densest collection of counting systems documented anywhere. And body tallies like the one used in Oksapmin are typical of central New Guinea, and particularly of its highlands. About 15 per cent of the counting systems of New Guinea are of this type: several dozen separate systems of counting gestures, all slightly different. They were used for traditional purposes such as bridal payments, and for counting valuables such as shells, pigs and other goods. In some languages, they were used only in special situations such as bridal wealth negotiations or festivals (it was usual for there to be an additional set of 'conversational numbers', as in Oksapmin); others used them constantly, with reports of individuals even counting collections of objects for the sheer pleasure of doing so.

Each of these systems uses gestures in a trajectory around the body, beginning with the fingers of one hand, proceeding up the arm and sometimes the head, and then back down the other arm to finish with the second set of fingers. Some are symmetrical, others not. Some have a central point like the nose or sternum, making the whole count an odd number; others lack a central point and count to an even total. The lower half of the body is not normally used. All the systems use or adapt the names of body parts as counting words, although a degree of mismatch has been reported in some cases and suffixes or other modifications can be involved, in order to distinguish the counting sequence from the ordinary names of the body parts. Many systems use a special prefix or suffix to distinguish the second half of the count from the first.

The length of the sequence varies widely. At least two body-tally systems count to forty-seven, and scattered reports suggest that systems existed enabling yet higher numbers to be counted.

Meanwhile, the shortest system recorded has just ten words and uses only one side of the body: possibly it has been truncated from a longer predecessor.

Where did this way of counting come from? With hindsight it seems natural, even inevitable, to extend counting on the fingers to counting on the body: but in fact, nowhere else in the world do people count using anything like the body tallies of New Guinea. Similar systems have been reported in the Torres Strait Islands and in southeastern Australia, but they are unknown further afield, and it appears they are an invention of the region, probably of the New Guinea highlands. It has been suggested that counts of this kind derive from using parts of the body – distances along the arm, for instance – to measure physical objects like strings of shells or ropes of rattan. Quite probably there was a single ancestral system at one time, subsequently transmitted and transformed around the region. An age around 9,000 years has been suggested, although other reconstructions are possible; certainly the body tallies underwent a complex pattern of changes during the centuries or millennia before written history.

The Oksapmin people were contacted in June 1938 by an Australian patrol trekking through the central New Guinea highlands, searching for minerals and for a site for a patrol station. An outpost of the Australian government was established in the area ten years later in Telefomin, about twelve days' trek from the Oksapmin. It was not until the 1950s that contact with patrols became at all regular, and only in 1962 were a patrol post and an airstrip built actually in Oksapmin territory; a mission station opened two years later. From the 1960s, the government and the mission opened schools in the Oksapmin area; these, and the possibility of employment at the patrol post, introduced the languages of Tok Pisin and English to the inhabitants.

The resulting cultural changes were far-reaching. Specifically for the Oksapmin counting system, increasing contact with outsiders, changing systems of government and the introduction of money wrought a remarkable series of alterations. Early payments by patrols to local people were in salt or matches, and there are reports of the use of cowrie shells as money in the early period of contact. By the late 1960s, Australian shillings had replaced other means of exchange (at a rate of about one shilling to one cowrie).

By this time, Oksapmin artefacts were being exported overseas: one writer reported seeing an Oksapmin arrow for sale in Brentano's Book Store in New York, for fifteen dollars. Employment at the airstrip, mission and patrol station became more regular, and opportunities arose for employment at plantations, and later mines, elsewhere on the island. A cash-based economy developed from the 1960s, in which labour, lumber and vegetables were sold.

The Australian currency was decimalised in 1966, with 100 cents to a dollar. Later, in the lead-up to its independence, Papua New Guinea issued a new currency with 100 *toea* equalling 1 *kina*, though the terms *shilling* and *pound* lingered in actual use long after their official withdrawal.

Meanwhile, from 1964 the mission ran a store at which it was possible to use currency to buy axes and knives, western-style clothing, matches, balls, mirrors . . . 'all types of useless and unnecessary items', in the words of one government officer. Wage labourers frequently invested their new wealth in goods such as rice and canned fish. Some opened trade stores of their own, and by 1980, there were more than a hundred such stores in the area, selling everything from food and clothing to soap and batteries.

All of these changes introduced new things to count, new structures for counting determined by the succession of currencies in use, and a new need to perform arithmetical operations in the context of – particularly – trade-store transactions. At the most basic level, this might mean counting coins in the same way other objects had traditionally been counted, matching them gesture-by-gesture with parts of the body. One pair of anthropologists report that 'in 1960, when Australian shillings and pounds were used in the Oksapmin area, a woman might establish a one-to-one correspondence between a succession of six body parts (in a conventional order) and 6 shillings and then indicate that the value of the shillings is *wrist*, the sixth body part in the series'. Addition problems might be solved by including two groups of coins in the count; subtraction by physically taking away certain coins and counting the group that was left.

Individuals near the beginning of their participation in a money economy, however, struggled when the coins were not present: struggled to count imaginary coins and reach correct answers.

Some people whose participation in the money economy (in 1980) was limited had not created keeping-track functions for body parts in arithmetical compositions. In their effort to add five coins to seven coins when the coins were not present to count, they began with the forearm (7), enumerating the elbow (8), biceps (9) and were unclear where to stop.

More experienced participants, however, created new strategies. Some used a subset of the body parts to keep track of the second set of coins to be counted: in other words they counted, say, from little finger up to thumb simultaneously with counting from forearm up to ear. Others simplified this procedure by deploying the *words* from little finger to thumb together with the *gestures* from forearm to ear. Finally, particularly among trade-store clerks themselves, there arose the so-called 'halved-body' procedure, in which one side of the body was used to 'hold' one number while the other side of the body held a second. By systematically trans-ferring units from one side of the body to the other, addition or subtraction could be carried out gesturally and the results translated into the now-required decimal form.

The most recent research on the Oksapmin counting gestures reveals a marked tendency to truncate the counting system at twenty, reflecting the importance of the 20-shilling pound and now the 20-*kina* note in currency transactions. The traditional '*fu*' meaning 'twenty-seven' or a complete count has been repurposed to reflect this: it no longer means twenty-seven but twenty, or in a money context 2 *kina* (that is, twenty 10-*toea* coins). From meaning two *kina* it is being extended by some speakers to mean two of anything else, and is even starting to appear as a suffix capable of doubling each of the number words from the traditional body-tally. The possibilities for novelty and change seem to be almost endless.

The New Guinea body-tally systems are a remarkable reminder of the range of possible ways of counting: a strategy that seems perfectly natural as an extension of finger-counting, but has arisen nowhere else on Earth. And the Oksapmin and others have been quick to innovate in their ways of counting in response to the need to count new things in new situations. Unknitting the link between body parts and words, they have given old words and gestures new meanings. It is a striking illustration of how adaptable counting systems can be.

Tonga: Counting leaves

The island of Tonga, 2007. A woman collects pandanus leaves ready to weave a traditional fine mat. She collects leaves from a special kind of pandanus, the *kie*, and processes them to obtain a light colour.

The kie leaves are cut while still green, and the sharp parts from both edges and from the middle rib are taken away. Most people also half them along the middle rib. Rolled into big slices, the leaves are then boiled for the whole night or day, after which they are softer, but still brown. At this point, they are then tied together into bundles and taken to the sea.

The count of leaves as bundled is a traditional one:

While their methods may differ for the small numbers – some count one by one, others in pairs – they make a bundle in the literal sense at the score (*tekau*): 20 leaves of pandanus is the number that is tied together in a bundle to be brought to the sea. These bundles or scores of leaves are then counted up to *tefuhi* ('10-scores') and beyond if applicable.

The process of leaf preparation continues:

close to the beach, a row of sticks is dug into the ground with one end, and the other end is connected with a string. The bundles are attached to this string, hanging down into the sea at high tide and exposed to the sun at low tide. After about ten days they are taken back, rinsed and rolled up to spirals, which are dried in the sun. When perfectly dry, they are smoothed, usually with a piece of metal, and then cut into tiny strips.

Pandanus leaves drying in Tonga.

The strips are finally woven into mats, which have a range of uses. Smaller mats can be worn around the waist or used as bedding. The most prestigious size of mat, about 2 metres wide and 10 metres long, would be a gift for a wedding or a funeral, or perhaps made for sale. Four or five thousand leaves would be needed for a mat of this size, and it might be worked on by a cooperative association of women, each providing an equal number of strips. The best mats become heirlooms; the collection of fine mats in the palace is sometimes described as the crown jewels of Tonga, and is displayed on state occasions. When Queen Salote Tupou III of Tonga received Queen Elizabeth II of the United Kingdom in December 1953, she wore a 600-year-old mat she had also worn at her own coronation.

The kingdom of Tonga consists of over 150 islands stretching over 800 kilometres of the Pacific, in an area around a third of the way from New Zealand to Hawaii. It lies on the western edge of the great scatter of islands known as Polynesia. Its climate is tropical, with warm humid summers, cool winters and much rainfall. The soil is fertile; so is the ocean. Fish and shellfish are traditionally important, as are vegetable products grown in a dense patchwork of fields. Agriculture, fishing and forestry provide much of the island's employment, its export earnings and its food. Few land animals, on the other hand, have ever travelled so far across the ocean, and animal food was traditionally for special occasions rather than daily consumption.

Tonga retained its political independence and its cultural heritage through the period of eighteenth- and nineteenth-century incursions, and remains a constitutional monarchy: the last remaining indigenous monarchy among the Pacific islands. The line of the Tuʻi Tonga chiefs can be traced back a thousand years.

About 100,000 people speak the Tongan language, which like many of its Polynesian siblings incorporates distinctive ways of counting. Its set of number words is a perfectly regular decimal one, with words for the numbers from 1 to 9 (*taha, ua, tolu, fā, nima* . . .) and for the powers of ten up to 100,000 (supplemented today with English words for 'million' and beyond). Multiplication and addition are used with near-perfect regularity to create terms for each number up to that limit.

Tongan possesses special words called noun classifiers, which accompany its number words and specify what kind of thing is being counted. Four categories of noun are distinguished: roughly people, animals, small or hand-made objects, and large or natural objects. The classifier words are also inflected to mark the grammatical distinction between dual and plural: between two things and more than two.

The system is further complicated by four separate ways of counting certain kinds of object, which involve some distinct terms for certain numbers. Sugar cane pieces are counted in pairs, with special words for 'pair', 'ten pairs' and 'twenty pairs'. Coconuts are counted in pairs initially, but for larger numbers they are counted in twenties, then in two-hundreds: there are special words for pairs, for twenties and pairs of twenties, and for two-hundreds and pairs of two-hundreds. Pieces of yam and pandanus leaves follow the same pattern of counting groups as coconuts, but their special words are different. Fish are counted in pairs and twenties, not in two-hundreds.

All of the special systems, in fact, combine the regular decimal system with pairing of objects and with grouping them in twenties. A result is that certain words are reused in different contexts. *Teau*, for instance, means a hundred ordinary things, but a hundred *pairs* of sugar-cane thatch, or a hundred *twenties* of coconuts, yam or fish. Indeed, there are even separate words in Tongan for 'counting in general' as against 'counting one by one': that is, not in pairs.

At first glance, the special counting systems are baffling in their complexity, as is the decision to count certain items in an unusual way at all. The items in question are natural subsistence products; what they have in common is not their importance or their abundance but their cultural significance, particularly in the context of prestige for gift-giving or feasting. These were products that on certain traditional occasions were collected in huge quantities, presented to and redistributed by chiefs and kings. Early descriptions suggest a frequent flow of food – in particular – into the chiefs' courts, and a corresponding regular redistribution on special occasions. Recently, birthdays, marriages and funerals have been marked in this way, with valuables exchanged. Presentation of certain foods to the king reportedly still used the traditional counting systems in the early twenty-first century.

One such presentation took place at the end of August 2004, when His Majesty, King Taufā'āhau Tupou IV, and several members of his family visited Pangai in Lifuka, Ha'apai, to preside over the Agricultural Show. The displays and presentations during this show bear a strong resemblance to the ceremony of first fruits, '*inasi*'. Such gifts usually comprise whole kava trees and sugar cane, yam, or giant taro, and senior pigs (*puaka toho*). Fish, turtles, octopus and giant clams can be part of the *fakapangai* as well, and sometimes baskets of fruits, woven mats from pandanus, and tapa mats (*ngatu*) are added.

While for the larger objects single pieces may suffice, . . . other objects have to be presented in bundles of 20. Particularly yam and giant taro are exclusively given and counted in *kau* (scores). Accordingly, the expected minimum is one score, and if more than that is given, it needs to be in multiples of the score.

Tongan is no outlier in these respects; similar systems are to be found in many of the nearly thirty languages of the Polynesian family, including Samoan, Tahitian, Maori and many others. Their actual number words are in many cases similar, reflecting their shared linguistic ancestry. Most use decimal number systems, and most also possess – or possessed – distinct counting systems for certain types of object, based on different counting units such as groups of two, four, ten or twenty. In general, the objects counted in special ways seem to have been those connected with traditional food production practices, with feasting or with ritual: resources that were both culturally important and abundant enough to be counted in large numbers. Fish were a frequent example; so were coconuts and tubers, and also the raw materials for making fabrics.

On Rennell, for instance, in the Solomon Islands, a Polynesian language is spoken whose counting includes special units for bananas, yam, breadfruit and taro. Bananas are counted in piles of four bunches; yam and breadfruit in pairs, with ten pairs to a basket; yam in piles of ten (or eight if they are large); topped taro in baskets usually of four pieces; taro stalks in bunches of five, twelve or twenty-two depending on type; taro tubers in bags. Coconuts in strings of ten nuts. Pandanus leaves in rolls of sixty or seventy-two. One report speaks of a collection of 7,600 piles of bananas: roughly 300,000 bananas in total when counted by these methods.

Examples could be multiplied. In Mangareva,

> breadfruit, pandanus leaves, agricultural tools, and sugar cane were counted in pairs (*tipau rua*); ripe breadfruit and octopus were counted in fours; and the first breadfruit and first caught octopuses of the season to be given as a tribute to the owner were counted in bunches of eight. All other things (including humans, mammals, or birds) were counted singly (*tipau tahi*).

In Samoan, there are special units for fish, young pigs and coconuts. And so on throughout the entire family of Polynesian languages. The same is true in the more distantly related languages of the Micronesian family: a decimal system in which many of the number words reflect a common origin with those of Polynesia is supplemented by – in some cases – a system of counting in pairs or other units.

Like many of the world's ways of counting, the Polynesian number words are under pressure. Words borrowed from other languages such as English have tended to enter the lexicon and replace terms for larger numbers. Experience with number symbols in a positional system results in number words being spoken not using the full, traditional decimal system but by naming the single digits; saying 'two seven four three' as the equivalent of 2,743, for instance. The more complex parts of the traditional systems are thus systematically lost: the special counting systems are no longer learned and people fluent in them become increasingly rare. Linguists report that those who know the old systems no longer agree about how they worked: 'Some do not apply or even remember the traditional systems at all, while others over-generalize some of the counting strategies.' Yet, like many human activities, the traditional Polynesian counting systems were well adapted to the purposes for which they were needed, and deserve to be understood and remembered, even if they may no longer be practised in the future.

———

The complexities of the Tongan counting systems and those of its linguistic neighbours and relatives beg the question of how things came to be this way. Archaeology bears witness to the spread of a group of seafarers out of southern China starting perhaps 6,000 years ago. The linguistic evidence – from the language family called Austronesian – tells the same story. These people spread first into Taiwan, then on to the Philippines; later to Malaysia, Indonesia, Madagascar, and east across the Pacific. They seem to have left Australia entirely alone (despite the perhaps confusing modern name 'Austronesian' for the language family), and to have settled only the more coastal parts of New Guinea, but they populated all the smaller islands. From perhaps 3,000 years ago they moved successively through Western and Central Polynesia and Remote Oceania. The limits of their voyaging were Easter Island, Hawaii and New Zealand. The last major archipelago was settled around seven hundred years ago. A consequence of this rapid spread across many islands is an extensive diversification of their languages: the Austronesian family accounts for about a fifth of living languages.

These were an adventurous people, riding the path of the winds to find tiny specks of land in a vast ocean. They farmed, traded, made the islands habitable by introducing plants such as taro, yam, sweet potato, banana, coconut, sugar cane, kava; they also brought with them some animals such as fowl, pigs and dogs. And they brought their words.

The proto-Austronesian number words, used by those who left the East Asian mainland 6,000 years ago, can be reconstructed with confidence (*a-sa, dusa, telu, sempat, lima . . .*) and it seems certain that their counting system was a regular decimal one. It is not clear how high they counted, but the first to enter Polynesia certainly had a word for 'one hundred', while proto-Polynesian itself had words up to ten thousand; possibly right up to a million. A decimal system with a large range is ancient in this part of the world, though the words for the very largest powers of ten seem to have developed independently on different islands according to need.

It seems certain that these large counting numbers arose in the context of resource collection and distribution at feasts and rituals: a cult of public generosity, in which the more a chief had to offer the gods and the people, the higher his status. 'On Lamotrek and Fais in the Central Carolines, for instance, large amounts of til fish were regularly distributed among the villages according to fixed proportions, and on Woleai, an instance of redistributing more than 12,000 coconuts locally during a funeral is documented.' In fact, it is not so much the accumulation of resources as their orderly redistribution that really requires an ability to count efficiently to high numbers.

As to the use of special counting systems to count certain items, their origin seems to be bound up with the use of classifier words in these languages. Many of the Micronesian languages and a few of the Polynesian ones, like Tongan, categorise nouns into as many as a few dozen classes. Compared with the four noun classes of Tongan, for instance, Samoan has fifteen and Trukese and Kiribati ninety, each with its own classifier word. This feature of the languages is certainly also an ancient one, present in the ancestral proto-Polynesian if not earlier.

Some of these classifier words not only specify the kind of thing being counted but also the unit in which it is counted: not bananas but bunches of bananas; not coconuts but strings of coconuts. It looks very much as though this was the origin of the special counting systems. In contexts where it was necessary to count large numbers of objects accurately, the task was simplified by adopting a classifier that specified you were counting not single items but groups of them: sometimes pairs, sometimes tens and sometimes twenties. This technique effectively extended the range of verbal counting without the need to devise any new number words.

———

Oceania has one of the longest and most complex and multi-layered histories of human settlement of any region outside Africa. Its people's traditional reliance on words to transmit ideas and culture makes it a showcase for what words can be and can do. The counting practices of Oceania are sophisticated, creative and perfectly suited to the purposes they served. They are the best evidence for what counting in words can achieve, and the unexpected range of ways it can achieve it. The Polynesian languages would not have run out of numbers when YouTube did. The Oksapmin have shown a remarkable inventiveness and flexibility in devising arithmetical procedures to use with their number gestures in contact with modern ideas about money and payment. And the tribes of Australia had all the number words they needed for their rich and diverse ways of life over tens of thousands of years.

Further around the globe, in the Americas, the picture is different in every respect. This pair of continents have the world's shortest period of human occupation, and perhaps its most diverse range of ways of counting.

8

Panorama: Counting in the Americas

Humans entered North America as the ice began to retreat: perhaps 14,000 years ago, perhaps more. They crossed on foot, likely aided by boats, over what is now the Bering Strait, from Siberia into Alaska.

They found an immense continent: a quarter of the habitable land on the planet, stretching nearly 16,000 kilometres from north to south. Over hundreds, perhaps thousands of years people spread down the coast and into the interior, populating North, Central and South America: the last major landmass to gain a permanent human population. They created a rich mosaic of cultures and languages.

By 1500 CE there were a thousand languages in the Americas. They do not group into a small number of large families, like the languages of Africa, Europe and Asia, but into a large number of small, unrelated families. It is the same with American counting practices. Almost everything to be found in the rest of the world can be found here: from simple counters and strung beads to elaborate systems of number words and number symbols, in a dense patchwork of unrelated – or only distantly related – practices.

So this chapter is something like a recapitulation. The last five chapters have followed counting around the world, from Africa and the Fertile Crescent, through Europe and India, to East Asia and the Pacific. This chapter follows it through the Americas, from Alaska to the Amazon. It follows the path of the first humans to set foot here, from north to south, and it tells not one story but many. For in this continent that is a microcosm of the cultural diversity of the planet, counting too shows just as much range, difference and surprise, as in the rest of the world put together.

Yup'ik: Counting games

Dug from the frozen soil of the Kuskokwim Bay in western Alaska, a bundle of thirty-eight matching wooden sticks: tally sticks used in a game, in all probability. Today they are stored at the museum at Quinhagak with many thousands of other finds from the same site.

Bundle of tally sticks from Nunalleq.

This find was made . . . in a very thin charred house floor under a wall. This house floor rests on natural [ground], so it represents one of the very first activities that took place on site. Close to the tally sticks were a large wooden spoon and the remains of a large clay pot. This and the evidence of burning make us think that this might be a cooking area – but what are the game pieces doing in the kitchen?

The American Arctic has long been the site of complex movements of population, driven at least in part by its ever-changing climate. It was around a thousand years ago that a group known today as the Thule culture appeared in the Bering Strait area of northern Alaska. They expanded down the western coast as far as the Kodiak archipelago and the Gulf of Alaska, and eastward across northern Canada and into Greenland, absorbing or replacing the existing populations. Their descendants live in the region today, and are known now as the Yup'ik people.

The village of Agaligmiut was occupied by the late fourteenth century; it stood in the Kuskokwim Bay, near both the ocean and the mouth of the river Arolik. At that time – the Little Ice Age – temperatures were a little colder than today, but this location was not the high Arctic: days were never shorter than five hours, and there were three months of summer when temperatures could push above 25° Celsius.

The whole Yukon–Kuskokwim region is a delta the size of a state: a wide marshy plain crossed by streams and ditches, dense with waterways. River mouths shift; the coast erodes. A single storm can change the shoreline out of recognition. It is a rich tundra landscape, and host to some of the planet's largest populations of water birds and fish. On certain days, the geese blot out the sun.

The village's site was chosen for its access to these natural riches: seals from the coast, salmon in the river and caribou on land. The main house was extended, remodelled and refloored on several occasions. First excavated deep into the soil, by its final phase in about the 1670s it had at least three separate entries, with long hallways and six or more side rooms.

The people's way of life was closely adapted to the resources of the site. They used wood and antler to make tools, weapons, game pieces, masks, kayak parts and arrows. Baskets were made of grass, clothing of skin and hide, other containers of fired clay. The Yup'ik harvested the fauna of land, river and sea with bows, harpoons and nets; they focused on the coast, but perhaps also made excursions to inland hunting and trapping camps for part of the year. Dogs pulled their sledges.

This was a broad, flexible lifestyle, capable of adapting to changes in the availability of one resource by leaning more heavily on another. The cooler period from the fourteenth century saw fewer fish, and the people turned to consuming an increased proportion of caribou, dog and seal meat: there is no evidence that they went hungry even in the coldest years. On the other hand, there is clear evidence of warfare across much of Alaska during the 1700s – possibly starting much earlier – and one suggestion for its cause is resource stress and perhaps competition for the most richly endowed sites, such as the Kuskokwim Bay. Raids, ambushes, and the destruction of whole villages took place. Agaligmiut itself was burned to the ground in a raid some time in the later years of the seventeenth century.

Today, the inhabitants of the area call the deserted site 'Nunalleq': the old village. Agaligmiut may in fact not be what the village was originally called: some say the element 'Agalik' refers to the local term for ashes.

A combination of ethnography and archaeological finds – like the set of tally sticks described above – reveals that this was a world

into which numbers and counting were richly woven. Games of
skill are mentioned often in descriptions of the Yup'ik lifestyle,
including darts games such as one observed in the 1890s. A wooden
block about 15 centimetres long was cut into a cylinder, flaring at
one end and pointed at the other. Planted in the ground, it formed
a circular target two or three inches across, with a deep hole cut
in its centre. Players gathered around it, cross-legged on the floor.
They passed around a small wooden dart, with which each player
tried to hit the hole in the top of the target. Nearby was 'a small
pile of short sticks, of uniform size, used as counters'. A successful
player would take one of these counting sticks, and would also
have a second try at hitting the target. When all the counters were
gone, the game ended, 'the players count up and the one having
the most counters [was] the winner'.

'Ordinarily this game is played by men, women, or children
merely for pastime', wrote one observer, 'but sometimes small
articles are staked upon the outcome. It is a source of much sport
to the players, who banter and laugh like school children at each
other's bad play.' Indeed, according to local tradition the burning
of Agaligmiut was an act of revenge provoked by an injury during
a game of darts.

Other dart-throwing games were played, including throwing at
an upright stick planted in the floor: ten wooden counting sticks
were used to keep score and determine which team had won. In
other games, the sticks could be both counters and part of the
play itself:

A bundle of from fifty to seventy-five small, squared, wooden splints,
about 4 inches long and a little larger than a match, are placed in a
small pile crosswise on the back of the player's outstretched right hand.
The player then removes his hand quickly and tries to grasp the falling
sticks between his thumb and fingers, still keeping the palm downward.
If one or more of the sticks fall to the ground it is a miss and the

next player tries. Every time a player succeeds in catching all of the falling sticks, he lays aside one of them as a counter until all are gone, when each player counts up and the one holding the greatest number is the winner.

The sticks could similarly be used for a game like jackstraws; they were placed in a heap and each player challenged to remove one without disturbing the others. Once again, the sticks functioned as counters: the player with the most sticks at the game's end was the winner. Of course, not all games were numerate in the same way: ball games involved keeping score but games of strength like wrestling or tug-of-war did not, and neither did speed trials such as foot- or kayak-racing.

Equally, not every use of counting was a game. Tallies of animals taken in the hunt were carved on ivory or whale tooth, either with simple straight marks or with incised representations of the animals in question: walrus, bear, geese. One beautiful example survives of a set of carved bird heads in a box, quite possibly another form of hunting tally.

More intimately still, some hunters were tattooed with – among many other marks and designs – tallies of the animals they had taken during their lives: caribou, or in communities where it was feasible to hunt them, whales. The hafts of harpoons were reportedly marked in the same way, to denote numbers of animals killed or, in one report from an Inuit community further north, the passage of time. In a tale about a group of Inuit lost on the ice:

'How long was it', I asked, 'from the time they were swept out to sea before they reach their homeland again?'. 'Ten summers and ten winters were the numbers of the notches that Comock cut on the handle of his harpoon', Nanook replied.

In the same way, ivory carvings of various designs were sometimes fitted with removable arms so that one could be inserted on each day of the week, forming a tally calendar.

———

The Yup'ik people counted in words, too. The language of Agaligmiut was the Central Yup'ik of the Yukon and Kuskokwim area. The group of five Yup'ik languages, of which this is one, were spoken across Alaska and the Russian Far East; its Inuit sibling is spoken further north in northern Alaska, Canada and Greenland. Together, the Yup'ik and Inuit languages – with the Aleut tongues of the Aleutian archipelago – make up one of the few really large language families of the North American continent, called Eskimo-Aleut.

The number words of Yup'ik correspond closely to a practice of gesture counting on the hands and feet: from left thumb to left little finger, continuing from the right little finger to the right thumb, and then similarly across the toes from right to left. The expression for 'five', *tallimat*, derives from the word for 'arm', *taliq*. That for ten, *qulit*, seems to mean 'top': that is, the digits of the top half of the body. The word for twenty, *yu-i'-nik*, meant 'a man is completed'.

Up to twenty, the count was in fives: albeit with some irregularities, so that expressions for nine and nineteen meant literally 'almost ten' and 'almost twenty'. Objects for counting seem regularly to have been grouped in heaps of five. Beyond that, the system worked in twenties, with the multiple of twenty given first, followed by the extra numbers. One situation where the larger number words were particularly important was when storing and counting fish.

Early in contact with outsiders, a word for 'thousand' was adopted from Russian – *tiisicaaq* – and its half used to denote five hundred, complicating the system further for larger numbers. During the twentieth century, the word forms also gradually changed, obscuring the relationship with body parts and on the whole making the system more regular, with a base of 5 for low numbers, and a base of 20 for larger ones. Central Yup'ik has over 25,000 speakers today, and its number words in this form are likely to survive into the future.

———

After its loss, Agaligmiut was rediscovered. Changes in the shoreline meant that the burned and buried village began to be eroded by the sea, its artefacts deposited in a trail along the beach: 'whittled bits of wood, sharpened stakes, and fragments of cut birch bark that were thinly distributed among more modern flotsam along the high tide line'. The site itself was emerging from the ground: 'Protruding from the dark soils were shaft fragments, pieces of bentwood bowls, and the trimmed timber supports of collapsed sod houses.' From 2009, it was dug by archaeologists with the consent and participation of the local people at Quinhagak. The cold, dry conditions meant that wooden, bone and antler artefacts – even woven grass – survived in staggering numbers. Some still bore traces of paint despite their centuries in the ground.

The set of thirty-eight tally sticks was excavated in July 2018. How were they used? There are many possibilities. For a game of darts, perhaps, or to report numbers of animals taken or keep track of the days. Perhaps they saw more than one type of use over the years. For they bear witness to an Arctic world in which counting was a part of everyday life; in which numbers, as at other times and places, were something to handle, grasp and manipulate.

Pomo: Counting costs

Take a clam shell (an empty one). Rub away the rough outer surface, use a stone to break the shell into rough discs, and chip off any pointed bits with a quartz blade. The discs that result are perhaps a centimetre across: you will be able to make up to forty from a single shell if you are careful.

Next, bore a hole in each disc with an implement much like a fire drill: a wooden shaft, tipped with quartz or flint, spun at speed by a taut string. This is the most specialised part of the work, and taboos surround those who perform it.

Then, group the beads, choosing a set of roughly the same size and threading half a dozen of them onto willow shoots. Roll them by hand on a sandstone slab lubricated with water, to take off any rough edges. (The combination of lime and water will wrinkle and seam the hands, so that a bead maker can be recognised at a glance.) A final polishing on deer skin is also possible.

Finally, thread the beads onto cords of wire grass, forming strings of up to two hundred. They will be stored in baskets or skin bags, or even buried to keep them safe.

The Pomo lived in what is now northern California, on the Pacific coast and perhaps as much as 150 kilometres inland. It was and is a varied terrain, with open country by the sea; there are mountains with belts of forest, sparsely wooded valleys, and lakes. Three thousand kilometres closer to the equator than the Yup'ik's Arctic homeland, it was rich in natural resources, something close to ideal for a hunter-gatherer lifestyle. The climate was on the whole very mild, and rainfall was reliable but not excessive. Drought and famine, though known, were rare, and the area in historic times supported upwards of 8,000 people. Estimates of the date of first human habitation in the area range as high as 9,000 years before the present.

Pomo shell money.

The name 'Pomo' groups together the speakers of seven related languages. They lived in small groups or bands, and their villages numbered several hundred, in addition to more or less temporary hunting and fishing camps. Their foods were acorn bread, nuts, seeds, bulbs and roots, as well as mammals, birds and – for those on the coast – molluscs. Trout and salmon were fished from the rivers and lakes with spear and net; clams and mussels were gathered from the sea. Deer and other large mammals were hunted with bow and arrow; rabbits and squirrels were trapped.

Much of their world was wooden: skirts were of shredded bark, rafts of logs. Houses were contructed from slabs of redwood bark and wood. Their woven baskets are famous to this day. For ornaments they had – as well as feathers and tattoos – beads.

Beads were being made in the California area soon after humans populated it. Over time, a number of mollusc species – abalone, clams, tusk shells and the so-called dwarf olives, among others – were used for their shells by different groups. It seems to have been the Chumash on the islands of the Santa Barbara Channel who first made shell beads in quantity, mainland tribes at first obtaining them by trade. Their olive-shell beads were in fact one of the first objects to be traded over long distances in North America; they appear in the Mojave Desert at very early dates. But, in time, inland groups became makers of shell beads too. At first they traded to obtain the raw materials:

> The Southeastern Pomo informant Wokox said that anciently clam shells . . . were obtained mostly through trade. Individuals seldom went for the shells for fear of trouble with the peoples along the long route. . . . The Coast Miwok of Bodega bay sold the shells to the Russian River people, who in turn sold them to the Wappo of the Middletown region, they to the Lake Miwok of Coyote valley, and the Lake Miwok sold them to the Southeastern Pomo.

But after white contact disrupted the Coast Miwok and reduced their numbers, certain Pomo groups began to make journeys, once or twice a year, to Bodega Bay to collect clam shell – as well as seaweed, salt and sea foods – themselves. The journey was one of well over 100 kilometres, and could take up to three or four days, involving most or all of the available adults in the community. The clams were dug using a stick to make a hole, from which the clam was pulled by hand; as much as a hundred pounds of shell could be carried home by each person.

By the later nineteenth century, the Pomo were the main suppliers of beads to central California, with the Yuki, Lake Miwok, Wappo and Wintu all receiving them through trade or gift. Bead making was now a specialised craft: most men made some beads, but some did little else. Some reports say that those who drilled the shells were subject to special taboos, rising early in the morning, working away from the house or abstaining from meat.

———

These Californians used their beads first as ornaments, notably as personal ornaments such as bracelets, pendants, necklaces and earrings. Descriptions from the historical period tell of wide wristlets made of fine beads, and elaborate complexes of necklaces and pendants dangling to the knees. There were beaded belts up to 20 centimetres wide and over a metre long. Nose sticks could have strings of abalone shells dangling from them. Beads were also used to decorate objects such as baskets and bags, in combination for instance with vivid red woodpecker feathers. For ceremonies, special quantities of bead decoration might be used, including special ornamented belts and hair nets. The figure of the 'bear doctor' wore bead strings and belts as part of his armour.

Beads – and beads in quantity – could mark prestige as well as decorate a person; it is reported that a Pomo leader could designate a successor by transferring a quantity of beads. In some parts of California, the inheritance of shell beads was treated with much care. Huge quantities could accumulate in some cases; photographs of Bear River people from the nineteenth century show individuals wearing ten metres or more of strung beads, numbering well over a thousand pieces of shell.

By the nineteenth century, it was a matter of frequent comment among the Pomo and their neighbours that the main use of beads was as a general medium of exchange. Their use was not restricted by class or sex, nor was it limited as to what things or services could be bought and sold. They were used to pay for food and objects, to pay doctors, shamans and singers; they were exchanged during marriage negotiations and to discharge blood debts. A deer cost 1,200 beads; a hand-made bow, 2,000 or more.

That is to say, beads had become money. California is one of the relatively few places in the world to have independently invented money – artefacts used to store and exchange value – and it is not perfectly clear when or under what conditions the innovation took place. Estimates have ranged from about eight hundred to nearly 2,000 years before the present, and there are several theories about the combination of factors that led to the development of money in this particular place and time. One view is that increasing population density, and possibly environmental variability, made it more than usually difficult for human groups to obtain the resources they needed either directly, or by barter exchange with their neighbours. Money, in this scenario, enabled resource transfers among groups that were not willing to incur debts to one another. By making exchange easier, it improved the overall efficiency with which resources could be gathered from the environment. The fact that beads already existed in large numbers – and were portable and durable – made them a natural choice for this new use.

Trade feasts provided a system for exchanging beads for a village's food surplus, and they illustrate the function money had in smoothing out variations in resources over time and place. A village with a surplus – of fish, say, or acorns – could invite others to a trade feast, setting the terms on which the surplus would be exchanged for shell money. The guests were allowed no choice as to the price paid. After the initial ceremonies,

> the host chief divided up the presented [beads] into strings of hundreds. These he placed upon the ground in a spot agreed upon, after the chiefs had arranged in council upon the amount of produce they were willing to give for each string . . . After this the several family representatives of the selling party went to their respective stores and each brought forth measures of produce to the value of one string. [Each giver] took the string of a hundred beads to which he was entitled.

The process continued as long as the food surplus – or the beads – held out, after which the chief of the guests divided the food among his people.

There were in fact two or even three standards of bead money in this system. The ordinary clam shell discs were the lowest unit of value; above them stood beads made from the heels of clam shells, about the same diameter as the common beads but longer, cylindrical. No more than four could be made from a single shell, and they were worth from twenty to forty of the disc beads. At the top of the scale stood beads of magnesite, a mineral that had to be mined and was itself sold between tribes. It had to be prepared in a similar way to shell, but inevitably involved more work: after roasting it was ground, bored, and finally baked and polished. A single piece could be valued at 800 shell beads. Red and yellow in colour, like shell it had a decorative function as well as being valued in exchange.

Many Californian tribes measured shell money rather than counting it. The central Miwok, for instance, measured strings by the *lua*, the distance from the nipple to the thumb and forefinger of an outstretched arm. Others used reference marks tattooed on their bodies to standardise the lengths. The Pomo, however, and others in central California, counted the beads using words. They gained, as a result, a reputation for their facility at counting, even into the tens of thousands.

In fact, the Pomo had a special set of counting words for dealing with beads. Their normal way of counting was in tens, but beads were counted in groups of four; two fours made (in the Eastern Pomo language) a *wedi*. Ten *wedi* were called 'one valuable' (sometimes a small tally stick was used to represent this number of beads). Eight 'valuables' were the next unit up: *dan ba'a*. Twenty *dan ba'a* is normally quoted as the limit of the system: *ethekai ba'a* or 12,800 beads. After this, the cycle would be repeated if there was a need to count even more beads.

———

For all the thousands and tens of thousands of beads that are reported to have existed, and been exchanged, counted and measured, fewer survived into the historical period than might have been expected. Beads were offered, sacrificed and destroyed on a variety of occasions: particularly large quantities were destroyed at funerals. While some Pomo groups allowed goods to be inherited, others insisted on the burning of half – or even all – of the possessions of the deceased: house and all, in some cases. Furthermore, funerary offerings were also made by relatives and friends, frequently in the form of beads.

Cremation, face down on a pyre in a shallow pit, was the normal way of disposing of a human body. Practice varied in detail, but the corpse could be wrapped in strings of shell, or practically buried beneath loose shell beads before burning. In some traditions, a final scattering of beads might cover the ashes after the cremation was over. As one informant put it, 'Two funerals in one month, week after week. That's where all my nice beads go.'

The use of bead money continued well into the period when the Pomo and their neighbours had access to American dollars. Rates of exchange varied around $5 or $10, sometimes less, per yard of strung beads.

After about the 1920s, the beads' function as money was lost, but beads did not cease to be made. Techniques evolved continuously; foot-powered grindstones were adopted, and drills powered by flywheels. Later, metal drills replaced the old stone ones. In the twenty-first century, some bead makers have adopted power tools: electric drills to perforate the beads and high-powered presses to shape them.

Like the Yup'ik far to the north, the Pomo wove counting into the texture of their lives, by adopting counting practices based on a combination of words and physical counters. They took the manufacture and exchange of counters to a high level of sophistication, arriving at a general-purpose money with only a handful of parallels elsewhere in the world.

Waxaklahun-Ubah-K'awil:
The long count

Oxwitik, 9.15.5.0.0 10 *Ahau* 8 *Ch'en* (22 July, 736 CE). Heir to three hundred years of the city's tradition, Waxaklahun-Ubah-K'awil, thirteenth ruler, divine lord, presides over the dedication of a new monument. Standing at the northern end of a 300-metre plaza that already contains one cycle of stelae, it will inaugurate a new series, binding heaven and earth, marking time and asserting the king's lordship over it.

The stone is 4 metres high, the glyphs of its inscription up to 30 centimetres wide. The inscription calls it a 'tree stone'. It specifies the date by saying that there are nine *baktuns*, fifteen *katuns*, five *tuns*, zero *uinals*, and zero *kin* completed; that the almanac day is 10 *Ahau* and that a certain god was lord of the night, while the position within the year is 8 *Ch'en*.

More than just a text, the inscription gives the date in a most elaborate visual form, each number represented by a humanoid figure and each unit of time by an animal that the person must carry: three birds, a toad, a monkey. The personified numbers employ tump lines, bands of cloth tied around the head, to bear the loads. They look like a row of porters in a caravan: merchants or pilgrims. Time, here, is a burden that the gods of number must laboriously carry through history; the stone is their resting place, where they can let their burdens fall.

Or some of them. Number five will be replaced by number six after this day, in time's great relay; but the personified numbers fifteen and nine will walk much further before their successors take over. The interlocking cycles of time are large indeed.

The Long Count at Copan.

The city of Oxwitik was part of the Mayan world: a network of highland and lowland city states linked by trade, by competition and conflict, by shared or related languages and by shared assumptions and rituals. Speakers of Mayan languages lived across the Yucatan Peninsula, and in what are now Guatemala, Belize and parts of Honduras and El Salvador. It was another abundant set of environments, and one of the handful of places in the world to have developed agriculture without outside influence. There were people living here perhaps 12,000 years ago; complex stratified societies by the start of the first century CE.

This is also one of the few places to have developed writing independently. Mesoamerica had at least three distinct traditions of logographs by the final centuries BCE; the relationship between them is poorly understood. By the third century CE, Mayan cities bore increasingly elaborate inscriptions on their monuments and buildings. The uses of text in this world were mainly for what might be called the propaganda or even the self-advertisement of the ruling elite: by contrast with early scripts in the Near East and East Asia, administrative and bookkeeping functions were strikingly absent. Not just carved on public monuments but painted on walls, fashioned in stucco or incised in pottery and even jewellery, the written messages highlighted key dates in the lives of rulers and cities: births, deaths, conquests, and the founding of cities. Writing that doubled as decoration was always capable of taking a wide range of forms, and artistic elaboration played a part in increasing the number of distinct glyphs: perhaps four hundred were in use at any one time, some standing for whole words and others for individual sounds.

Tantalising archaeological finds suggest there were also books, made of bark or animal hide folded concertina-fashion. They are depicted on pottery, and are occasionally found dissolved and unreadable in graves, but none survive in a legible state from the classic Maya period.

As well as symbols for words, the Maya had symbols for numbers, and they are justly famous for their seeming obsession with complex calendrical information, witnessed in many of their inscriptions. Many cultures have counted days, but few if any have counted them with the panache and sophistication of the classical Maya.

The Mayan languages had distinct words for the numbers from
1 to 12, and they formed the subsequent -teens as compounds with
'ten'. In the related modern language of Yucatec, numbers from
1 to 5 are *hun, caa, ox, can, hoo*, while 10 is *lahun* and 13 – for
instance – *oxlahun*. From twenty upwards, the counting words
worked purely in base 20, and it is likely that there were
words for both 400 and 8,000. Underlying meanings relating to
'person' for 20 and 'sack' for 8,000 have been reported, reflecting
the full count of fingers and toes and the contents of a sack of
cacao beans. Beans, or small stones, may have been used on the
Mayan counting boards described by much later sources in Spanish,
but no archaeological evidence for them has been found.

And of course, there were number symbols as part of the Mayan
script. A dot meant 1, and could be repeated up to four times. A
bar meant 5 and was repeated up to three times; bars and dots in
combination thus counted as high as 19, the limit of this part of
the system. (As with other counting technologies elsewhere in the
world, no element in the system was repeated beyond the subitising
limit of four times.) Like other glyphs, the number signs could be
used as syllables, and they also appear in the names of divinities
– 'nine strides', 'ten sky'. Their use as syllables, indeed, resulted in
Waxaklahun-Ubah-K'awil's name being read for some time by
archaeologists as '18 rabbit'.

Their main use, though, was in the Mayan calendar. Or rather
calendars, for there were several, and they interlocked to evoke the
vastness of time to an unparalleled degree. There were twenty names
for days: *imix, ik, akbal* . . . These were crossed with the numbers
from 1 to 13 to give an almanac cycle of 260 days, each specified
by a unique combination of name and number. This system was
already present in pre-Mayan texts of the first millennium BCE,
and possible explanations for its unusual length range across astronomical phenomena and the duration of a human pregnancy, or
possibly the length of the Mesoamerican agricultural season.

Next, and quite independently of the 260-day calendar, the year was built up from eighteen named months, each with twenty days, plus a sub-month of five days, making a total of 365 days. This, too, was an ancient system, and it provided a second way to name any given day within its cycle.

Third and finally, a neater year of just eighteen months – 360 days – known as the *tun*, was used as the basis for specifying dates in the so-called 'long count' calendar. *Tun* were grouped in twenties (*katuun*); *katuun* were grouped in twenties too (*baktun*). From a start date placed in the mythic past, the Maya could specify any day by giving the number of *baktun*, *katuun*, *tun*, months and days that had passed. This system was quite specific to the Maya, and indeed strongly associated with the classical period from roughly 300 to 900 CE (in Mayan terms, from 8 *baktun* 12 *tun* to 10 *baktun* 4 *tun*).

Many Mayan inscriptions began by naming the date in all three of these ways; like that on the stela at Oxwitik, dedicated on a day identified as long count 9.15.5.0.0, almanac day 10 *Ahau* and year day 8 *Ch'en*. With three different cycles in play, of 260, 360 and 365 days, there were many different possiblities for anniversaries and commemorations. It turns out, for instance, that fifty-two long years make exactly seventy-three repetitions of the almanac cycle, and this so-called calendar round was carefully observed as an anniversary. Meanwhile, it was of particular interest if the almanac day matched that of a previous event being commemorated. Furthermore, new years and the ends of *katuun* and *baktun* were treated as special anniversaries (very roughly like decades and centuries in the Western decimal count of years); half and quarter *katuun* were also observed in some cities.

When recording an end-of-cycle date, the final two or three number 'slots' in the long count date were necessarily empty, and the Maya used their sign for the adjective 'no' to meet the need: 'there were 9 *baktun*, 15 *katun*, 5 *tun*, no months and no days'. The sign was therefore functioning as something like a zero in the dating system: one of the world's rather few independent inventions of a symbol for zero.

———

There was something exuberant about the Mayan counts of days within the various cycles, and in many of the Mayan inscriptions the abundance of their numerical information was augmented by providing still more information beyond the three basic counts. Some gave the day's position in a nine-day cycle of 'lords of the night'; some specified the age of the moon and the length of the current lunar month: either twenty-eight or twenty-nine days in the conventionalised calendar for such things. Some mentioned yet another calendar, totalling 819 days. Others referred to dates in the mythic past that lay before the zero date of the long count: that is, during a putative previous long count which had totalled thirteen *baktun*. Others again evoked the far distant future, using rare glyphs for multiples like *pictun* (8,000 *tun*), *calabtun* (160,000 *tun*), *kinchiltun* (3,200,000 *tun*), and *alautun* (64,000,000,000 *tun*). One stela continued this series of ever-larger time units to twenty-four iterations: a count of years that would fill thirty-one figures in a decimal representation.

It is thanks to this near-obsession with dedicating and dating, commemorating and counting, that the history of sites like Oxwitik – now called Copán – is well understood today. Nestled in a valley at about 900 metres and watered by the Copán river, it occupied a pocket of rich agricultural land and relied heavily on maize farming. Occupied by non-Maya people as early as 1400 BCE, it was taken over in 426 CE by a new, Mayan ruler, installed by the nearby city of Tikal. This outsider king busily erected buildings and monuments, and founded a dynasty that would last for seventeen generations, eventually ruling over a population of perhaps 20,000.

His successors created temples, plazas, open-air altars, ceremonial

ball courts and a complex of royal residences at Oxwitik, a group
of monumental buildings whose impact was immediate and spec-
tacular. They diverted the river, and they commissioned some of
the richest and most distinctive sculpture of the Mayan period. As
in other Mayan cities, society became steadily more stratified and
production more specialised, with plants, animals, trees and mineral
resources all exploited and traded across the region. At different
times seashells, greenstone beads, cacao beans and copper bells
were all used as money. The city's pottery became a prized export.

Waxaklahun-Ubah-K'awil (his name may mean 'Eighteen are
the Images of K'awil') took the throne in July 695 (specifically,
9.13.3.6.8, 7 *Ahau*, 1 *Mol*). Over several decades, he built temples,
a tomb for his predecessor and father, and a huge court for the
ceremonial Mayan ball game. And he commissioned a remarkable
series of stelae for the main plaza of Oxwitik. Each depicted the
king performing a ritual, perhaps engaged in trance, accompanied
by supernatural beings; each was adorned with a text giving its
date and the dates of the acts of ancestors or divinities which it
echoed and commemorated. Together, they turned the great plaza
into something like a sacred space.

A first series of six stelae was completed by 9.15.0.0.0 (22 August
731); five years later, the king began a new set. The new stela at the
northern end of the plaza bore sculpture of fantastic beauty and
complexity, portraying the king masked and accoutred for ritual
and flanked by eight manifestations of a single god. The side bearing
the inscription gave the date using not the dot-and-bar number
symbols of many Mayan inscriptions, but an alternative system. Each
number up to 12 (including zero) was associated with one of the
gods who were used to name the days in the almanac. For numbers
between 13 and 19, the ten and the smaller number were combined
into a single figure, echoing the structure of the Mayan number
words.

This list of gods (specifically, the set of pictures of their heads) was a common way of writing numbers, second only to the bar-and-dot symbols. It was capable of huge artistic elaboration, and on occasion the gods could be depicted as complete anthropomorphic figures; this was how they appeared on the new stela. Combined with the set of zoomorphic divinities associated with the different units of time, the result was a complex, naturalistic scene of number-gods carrying time units. One of the pinnacles of Mayan sculpture, it is among the most elaborate and beautiful of all visions of counting.

If the king was the owner of time and the controller of ritual, he was also the monopolistic wielder of force, both in the form of the labour used to build temples, monuments and the rest, and in the form of war waged on his neighbours, in a continual jockeying for control of resources, trade, tribute and prestige. The Mayan kingdoms were never unified into a single state, and they spent much of their energy in strife against one another.

Fifty kilometres to the north of Oxwitik lay Quiringa, apparently a client or a tributary city. Its new king was inaugurated in 724 CE under the supervision of Waxaklahun-Ubah-K'awil. Over the following decade, something went wrong with this relationship: aggression on one part or the other, or a desire for independence by the ruler of Quiringa. Possibly with help from another larger city in the region, king K'ak' Tiliw Chan Yopaat of Quiringa succeeded in capturing Waxaklahun-Ubah-K'awil. On 3 May 738, he beheaded him.

Naturally, there are monuments at both cities giving the date. At Oxwitik the description states that the king's 'breath expired in war' and lamented that the city now had 'no pyramid, no altar, no cave': that building work was at a standstill, as was ritual access to the underworld. The defeat was felt almost as a cosmic disaster.

There would be four more kings at Oxwitik, but the ninth

century was a period of decline across the whole Maya world, for reasons that probably included both internal tensions and over-exploitation of the environment. Populations declined, dynasties collapsed, cities were abandoned. Oxwitik never really recovered from its disaster, and the last king was dead by 820. Buildings crumbled, and the river channel shifted, cutting a swathe through the ruins. By the tenth century the whole valley was abandoned.

A few Mayan books from the tenth to the twelfth century survive, giving a vivid sense that astronomical and calendrical information was still being transmitted, centuries after the peak of Maya culture. The 260-day almanac cycle was retained for centuries (it is reportedly still in use today), but the long count was gradually abandoned. Dates came to be given in a simplified form, stating the number of *katun* but omitting the largest unit, the *baktun*. This was a system that would repeat every 256 years.

The bar-and-dot numerals were also abandoned, although further south in the Andes the Mixtec and Aztec cultures would adopt related systems and transform them for their own uses. One consequence of these changes is a lasting uncertainty about how exactly the classical Mayan calendar correlated with calendars elsewhere in the world: which day of the Western calendar corresponded to its start. It was around 11 August 3114 BCE, but corrections of one or two days have been proposed from time to time, and certainty is elusive.

The Mayan people described and counted longer spans of time than any culture before or since, and their sculpture endures as a monument to their skill and imagination. Theirs was a world pervaded by numbers as dates, timescales, commemorations; even as gods. On the great stela at Okwitik they left a memorial of their king, their city, and one of the supreme expressions of the power and value of counting.

Pirahã: Lost count

Xagiopai, in lowland Amazonia, on the banks of the Maici River, 2004.

A member of the Pirahã tribe faces an anthropologist. The anthropologist sets out a row of batteries on the ground and asks his interlocutor to make another row matching it one-to-one.

What will happen?

Deep in South America, 10,000 kilometres from Agaligmiut and 3,000 from Copán, the Pirahã live – today – beside a tributary of a tributary of the Amazon. They number about four hundred, and they are hunter-gatherers. Their villages – each of about ten or fifteen adults – have some contact with Spanish-speaking traders, and they have a history of more than two hundred years of contact with Brazilians as well as their neighbours the Kawahiv. But they have rejected assimilation into the mainstream culture of Brazil, and remain monolingual in the Pirahã language. Linguist Daniel L. Everett, who spent over six years living with the Pirahã, writes:

> The Pirahã are some of the brightest, pleasantest, most fun-loving people that I know. The absence of formal fiction, myths, etc., does

not mean that they do not or cannot joke or lie, both of which they particularly enjoy doing at my expense, always good-naturedly.

Their language is the last surviving member of its family. It is famous for a number of unusual features, which threaten to break several otherwise attractive generalisations about human languages. In some ways, it is a complex language, with intricate verbal morphology and a rich five-way system of syllable types distinguished by weight. On the other hand, its set of sounds is one of the smallest in the world; Pirahã women use just seven consonants and three vowels (men have one more consonant). It is the only documented language with no terms for colours; it has remarkably few words for time, and it possesses the simplest set of pronouns known (which appear in any case to be borrowed from another language). It has no perfect tense: no way for a verb to express specifically the completion of an action described.

The cultural factor thought to lie behind these various features of their language is that the Pirahã people communicate only about subjects within their immediate experience: things that have been seen or recounted by someone now living. They completely avoid talking about the abstract or the second-hand. Everett's example:

> 'I prefer whole animals to portions of animals'. (Literally 'I desire [a] whole animal[s], not piece[s].') Sentences like this one cannot be uttered acceptably in the absence of a particular pair of animals or instructions about a specific animal to a specific hunter. In other words, when such sentences are used, they are describing specific experiences, not generalizing across experiences.

Other consequences of this are that abstract structures like long genealogies or complex kinship relationships are absent from the Pirahã culture. Kinship terms refer only to relatives a person knows, never to those who died before that person was born (Everett 'could not find anyone who could give the names of his/her great-grand-parents, and very few could remember the names of all four grandparents'). There are, furthermore, no fiction, no creation stories, no myths nor any other stories about the ancient past.

And there is no counting.

—·—

First, the Pirahã language has no distinction between singular and plural. No part of speech is marked to show a distinction between 'one' and 'many', so – unlike in most languages – you don't have to answer the one/many question every time you utter a sentence: indeed, you couldn't if you wanted to.

Second, the Pirahã have no counting routines, no counting practices, no habit of putting objects in a sequence. Counting with fingers or with counters is unknown here. And Pirahã has no counting words. Early investigation reported the terms *hói, hoí* and *bá a gi so* as meaning something like 'one–two–many'. But further investigation has revealed that these terms are not part of a counting sequence, and they are not number words at all; they mean respectively a 'small size or amount', a 'somewhat larger size or amount', and 'cause to come together' or 'many'. None is used consistently to refer to a specific number. The language has no other terms for quantity such as 'all', 'every', 'most', 'each' or 'some', nor words for items in a sequence such as 'first' or 'last'. *Hói* often denotes a single one of something, but it doesn't have to; it can mean as many as six. Similarly, *hoí* can denote groups from two up to ten objects, depending on context.

During the Brazil nut season, there are regular visits to the Pirahã villages by river boats, a contact that has probably lasted more than two hundred years, with Pirahã men collecting nuts and storing them to trade. But the trade is not numerate. The Pirahã, despite this long contact, have not learned the Portuguese number words and barter with – reportedly – little regard for the quantity of goods involved. 'Someone can ask for an entire roll of hard tobacco in exchange for a small sack of nuts or a small piece of tobacco for a large sack.' 'In this "trade relationship" there is no evidence whatsoever of quantification or counting or learning of the basis of trade values.'

———

Anthropologist Peter Gordon, in 2004, involved members of the Pirahã community in a series of experimental tasks, to learn more about what quantity did and did not look like in their world:

> I sat across from the participant and with a stick dividing my side from theirs, I presented an array of objects on my side of the stick . . . and they responded by placing a linear array of AA batteries (5.0cm by 1.4cm) on their side of the table.

He emphasised, when describing this scene, that the Pirahã clearly understood the tasks and were trying hard at them.

Matching a small number of items – two or three – often produced a one-to-one match or something very close to it. Matching larger numbers of items produced more approximate representations, with the numerical match falling off as the numbers became larger. For the Pirahã, it seems, a set of nine objects and a set of ten are distinguishable, but only just: perhaps three-quarters of the time.

From simple rows of batteries, Gordon moved on to 'clusters of nuts matched to the battery line, orthogonal matching of battery lines, matching of battery lines that were unevenly spaced, and copying lines on a drawing'. A final task involved watching nuts placed in a can and then taken back out one by one, with the experimenter asking at each stage whether there were any nuts left in the can.

For these more complex tasks, some of which involved transferring counts across time or space, the representations produced by the Pirahã were still more approximate, with exact matches produced rarely or not at all for the larger sets of objects. The range of their representations became wider as the sets became larger, showing the classic characteristic, in fact, of the approximate number sense. A colleague later performed related experiments, presenting the Pirahã with two rows of objects and asking if they were equal in number. The results confirmed still more directly that small numerical differences simply were not perceived – were not experienced. Indeed, with no way to perceive or to name exact quantities, it appeared that *exact numbers* simply were not concepts for the Pirahã.

Like all humans, the Pirahã experience number approximately, with differences among adjacent numbers less and less perceptible as the numbers become larger. Unlike nearly all adult humans, however, they do not overlay this perception with any other way of counting.

The Pirahã and their remarkable language caused a good deal of surprise when first reported in journals of anthropology and linguistics. Debate will doubtless continue about what exactly certain words in the Pirahã language mean or do not mean, but the fact remains that no one who has visited the tribe has ever claimed to see them spontaneously counting anything. Two further teams of anthropologists repeated certain of the number-matching tasks. One thing that emerged from this contact was a desire of some of the Pirahã to learn counting words. Everett responded to their request to learn to count in Portuguese, and reported that after

eight months of daily effort none had learned to count to ten. A
second attempt was more successful. A researcher who spent several
months in the Pirahã village of Xagiopai trained its inhabitants in
one-to-one matching tasks and invented Pirahã-like words for
numbers up to ten. The result was – in that village alone – a
significantly more exact set of responses to some matching tasks
by the end of the period. This certainly confirmed that the Pirahã
are capable of learning to count – that there is no question of an
inherent inability here – but it does not seem to change the fact
that, in their language and culture as originally documented,
number is experienced only as approximate. It is not certain quite
how rare this view of quantity is among humans: it has been
claimed of a number of other, mainly Amazonian groups that their
languages contain no number words, but none has received
anything like the systematic attention paid to the Pirahã.

———

The Pirahã represent a remarkable case within the range of possibilities for human language and culture, and a quite astonishing
branch on the story of human counting. The knowledge that there
is a living culture with no counting at all is a salutary reminder
that counting – in the sense of repeated attention plus a way to
keep track of it – is not inevitable for human beings. It is not clear
just what sequence of events led to the existence of an Amazonian
tribe with no counting practices, but the prevalence of counting
words across the Americas makes it scarcely credible that the Pirahã
somehow represent a line of humans that never learned to count
in the first place. It seems all but certain that they are descended
instead from ancestors who did count.

The latter possibility is hinted at by the fact that Mura, a now-
extinct dialect closely related to Pirahã, possessed some of the
cultural artefacts Pirahã does not: fables and legends and other

fiction about the distant past. Perhaps it had number words as well, or a common ancestor language did so.

If the Pirahã are indeed descended from people who counted, the situation is quite a dizzying one. Not only is counting a cultural artefact, an invention that – it turns out – not every society possesses, since not every society lives a lifestyle that needs it. But counting is capable of extinction if the need for it passes away. One of the branches on the story of counting ends – for the moment at least – in a world without counting. One of the possible futures for any human culture is indeed a world *after counting*, in which the skill – for all its range, power and great age – has been forgotten. In which counting words have fallen silent, counting gestures have been forgotten, tallies are no longer made and beads are once more just beads.

Conclusion

So many hands and bodies and voices and minds have counted. Mentally, manually, with materials, with words, with marks, with symbols, with machines. They have counted people, possessions, food items, friends, enemies, votes, years, days, eggs, trees, coins. They have counted to pay their taxes, to ensure their survival, to administer their cities and their businesses, to measure their self-worth, to commemorate their ancestors, to buy or sell. They have counted for fun. There is always a strange alchemy to counting, which restlessly transforms one thing into another: days into tally marks; people into counters; books into magnetic tapes.

This book has described a few of those processes. There have been thousands more, in all the thousands of languages living and dead, in all the thousands of cultures and hundreds of scripts. No two are alike. The real alchemy, perhaps, is in turning all of these processes into a single thing, and calling it 'counting'.

Counting is – this book has assumed – repeated attention to things or events, plus a way of keeping track. The different things to which attention might be paid, together with the manifold ramifications of 'ways of keeping track', have produced an almost inconceivable complexity and diversity of counting processes at different times and places.

Counting builds on innate abilities: the ability to estimate the relative sizes of groups of objects, and (possibly) an additional ability to recognise at a glance – to 'subitise' – sets of one, two, three or four. At least the first of those abilities is shared widely among the primate lineage, as well as with some other mammals and with birds; possibly with other animal groups as well. But no animal can count precisely. No animal species has been observed spontaneously using counters or tallies in the wild; none can learn more than the first few in a set of number words or number symbols. Counting uses regions of the brain innately specialised for the approximate number sense. Depending on what device for keeping track is involved, it engages other brain regions as well, including those that process spoken or written language, or those that handle fine motor control.

The ancient Stone Age environment contained several structures that could be used for the keeping-track function: structures involving a stable sequence of objects or actions. The marks produced by repeatedly striking stone on bone are found up to 3.2 million years ago; examples interpretable as tallies start to appear perhaps 70,000 years ago. Beads on strings also appear in the East African archaeology by at least 70,000 years ago, very possibly earlier. Extended or folded fingers first appear in archaeological evidence just 27,000 years ago, in Europe, but it is overwhelmingly likely that people counted on their fingers in Africa tens of thousands of years before. Vocalisations leave no archaeological trace, but there were very possibly conventionalised sets of sounds accompanying repeated actions early in the story of language; it is impossible to say when they were incorporated into speech or transformed into cardinal – ordinal, distributive, frequentative, and so on – numbers. Very nearly every documented language has some number words, and considering what they have in common enables (only) a few steps towards imagining what the earliest number words may have been like.

Number symbols – potentially a way of counting, but of most use as a way of recording and communicating the outcome of a count – appeared alongside the earliest notations recording spoken words, in contexts from the Near East, East Asia and Central America. There are constraints on the structure a set of visual symbols can have if it is to encode numbers unambiguously: each of the main possibilities has been used successfully and stably over long periods of time and large areas of space. Around a hundred distinct sets of number symbols are documented from around the world.

On the basis of these origins, counting has developed in a vast variety of ways at different times and places, its roots supporting many, many branches. Humans faced with different things to count and different reasons for counting have been hugely creative in finding new ways to count, and in adapting the old ways. Counting practices have always been devised in response to particular needs at particular times and places; they have always emerged in the course of specific interactions between people and their world. Some have been transmitted to neighbouring cultures and subsequent generations, but more have been short-lived, persisting only briefly before being transformed again or replaced by new experiments and inventions. Feedback has been an important force: once people gained new technology – a new way of counting – they invariably found new things that they could do with it; perhaps things they never previously knew they wanted or needed to do. And those new functions sometimes, in turn, drove new innovations in ways of counting.

In a lot of the situations where people have wanted to count, they have also wanted to do at least simple arithmetic, so there has been an overlap between the technologies used to count and those used to calculate. A lot of devices for keeping track have been adapted, more or less drastically, to provide ways of doing arithmetic. At least sometimes, considerations of their efficiency and effectiveness for calculation have affected the choice of which technology to use for counting itself. The shift from counting board and Roman numerals to paper and Arabic numerals in medieval Europe is one such case;

the rise of the *suanpan* and *soroban* in East Asia is another. On the other hand, there have been plenty of times and places when counting and calculation used separate technologies.

Over time, manual and verbal ways of counting have replaced one another; spoken and written numbers have gained or lost in importance, more concrete or more purely mental ways of counting and calculating have advanced and retreated. The accidents of history have given prominence to one way of counting, then to another. The simplest ways of counting (fingers, beads) have never really vanished; neither has the innate ability to estimate. Nor have the limitations of the subitising system, and their consequences for the devices people build and the notations they devise.

The story of counting is a dense tree, with several roots and nearly an infinity of branches. Its end points – as the tree stands today – include elaborate routines with unmarked counters on marked surfaces, highly developed sets of number words, successful – and successfully exported – sets of number symbols, and electrical machines whose internal states are, for some purposes, taken to represent numbers. They include smaller sets of words, sometimes accompanied by gestures, tally sticks and written representations of number words. They include a world in which there is no counting whatever.

What will counting look like in the future? The tree will continue to grow and change. Some branches will burgeon with new growth; others, not so much. Over time, those patterns will themselves change: human ingenuity is most certainly not finished with counting yet. Almost nothing can be said with confidence about the detailed forms that counting will take in the future. But they will continue to be manifold, various, surprising and fascinating: there will always be much that is new in the story of counting, as well as much that is old.

Acknowledgements

I am grateful to the late Felicity Bryan, whose interest and enthusiasm helped to shape this project in its earliest stages. My current agent Carrie Plitt provided valuable help during the later stages of work on it.

My thanks to Arabella Pike and her team at William Collins, including Sam Harding and Alex Gingell.

Christopher Hollings read the book in draft, as did my parents Moira and Tony Wardhaugh; their comments and suggestions were of great value.

As always my greatest debt is to my wife Jessica and our three sons, whose support, advice and practical assistance at all stages of this project were invaluable.

Image credits

p.20: kristof lauwers / Alamy Stock Photo.

p.27: imageBROKER.com GmbH & Co. KG / Alamy Stock Photo.

p.33: Heritage Image Partnership Ltd / Alamy Stock Photo.

p.41: agefotostock / Alamy Stock Photo.

p.45: Musée d'Aquitaine, Bordeaux, Gironde. Japhotos / Alamy Stock Photo.

p.47: SOPA Images Limited / Alamy Stock Photo.

p.75: Metropolitan Museum of Art, 1988.433.2. Public domain.

p.89. Nineveh, South-West Palace, Iraq; British Museum. Lanmas / Alamy Stock Photo.

p.93: British Museum, EA 10529 (recto). ©The Trustees of the British Museum. All rights reserved.

p.103: Wikimedia Commons. CC BY-SA 2.5.

p.113: British Library, Royal 13 A. XI, f.33v. By permission of the British Library.

p.122: The Morgan Library and Museum, MS M.240, fol. 8r. incamerastock / Alamy Stock Photo.

p 125: H. de La Tour, *Catalogue de la collection Rouyer* (Paris 1899), plate 6, figure 1. Public domain.

p.126: Jacob Köbel, *Ain New geordnet Rechen biechlin auf den linien mit Rechen pfeningen*, title page. München, Bayerische Staatsbibliothek, 4 Math.p. 175 va. CC BY-NC-SA 4.0.

p.128: Deutsches Museum, 1987-178. CC BY-SA 4.0.

Notes on sources

This is a work of synthesis, based largely on survey articles and outlines of the handbook/companion type, supplemented for the detailed case studies by publications discussing and/or reproducing specific sources. What follows is a list of the principal sources used for each section of the text, including the sources of all direct quotations; it does not list exhaustively the minor or supplementary sources consulted, nor attempt to specify the authorities standing behind every assertion.

A high proportion of the topics addressed have been the subject of disagreement among specialists, and in some cases that disagreement is ongoing. In such cases I have attempted to follow the weight of recent opinion; that is, I have not intentionally presented outdated or minority interpretations. Alternative views are in certain cases mentioned within these notes.

Confusions and errors in the text are naturally my responsibility, not that of the scholars cited below, to whom grateful acknowledgement is made and from whom indulgence is craved. In a project of this kind, reach necessarily exceeds grasp, perhaps by some distance; one of its premises is that the task of synthesis nevertheless has value.

ABBREVIATIONS

BNP   Cancik, Hubert, *et al.*, eds., *Brill's New Pauly* (Leiden, 2012).

CEWAL   Woodard, Roger D., *The Cambridge Encyclopedia of the World's Ancient Languages* (Cambridge, 2004).

CWP   Bahn, Paul and Colin Renfrew, eds., *The Cambridge World Prehistory* (Cambridge, 2014).

EHAS   Rashed, Roshdi, ed., *Encyclopedia of the History of Arabic Science, Volume 2: Mathematics and the Physical Sciences* (New York, 1996).

HMC   Campbell, Jamie I. D., *The Handbook of Mathematical Cognition* (New York, 2005).

OCD   Hornblower, Simon and Antony Spawforth, *The Oxford Classical Dictionary* (3rd edition, Oxford, 2005).

OHHM   Robson, Eleanor and Jacqueline Stedall, eds., *Oxford Handbook of the History of Mathematics* (Oxford, 2009).

OHNC   Kadosh, Roi Cohen and Ann Dowker, eds., *The Oxford Handbook of Numerical Cognition* (Oxford, 2015).

WALS   Dryer, M. S. and M. Haspelmath, eds., *The world atlas of language structures online*, Munich: Max Planck Digital Library, <http://wals.info>.

WWS   Daniels, Peter T. and William Bright, *The World's Writing Systems* (Oxford, 1996).

Introduction: What is counting?

The definition quoted from Leibniz appears in the *Confessio philosophi*; see Wedell 2015, citing Leibniz 2005, p. 102.

It is not feasible to list here the large number of public-facing accounts of number and its history, many of which intersect with the concerns of this book. As well as those mentioned below for specific points, I have found the following of particular help in arriving at a sense of the general shape of the story (or stories) of counting: Scriba and Ellis 1969, Flegg 1983, Crump 1990, Ifrah 1998, Chrisomalis 2020.

1 Number sense before counting

The literature on innate number sense(s) is very large, and new research is constantly being published. Helpful survey collections are HMC, OHNC and Butterworth *et al.* 2018. Summaries of the animal evidence therein are Brannon's chapter in HMC and Agrillo's in OHNC. Among the many papers cited there, I have used specific evidence from Ditz and Nieder 2016 (on crows), Benson-Amram *et al.* 2017 (on social carnivores), Hauser *et al.* 2000 (p. 829 for the macaques in the Cayo Santiago), Uller *et al.* 2001, Cantlon and Brannon 2007a, 2007b, and Beran *et al.* in OHNC (on primates).

The neuroimaging evidence is discussed in Nieder's paper in OHNC, Amalric and Dehaene 2017, Hyde 2011, Anobile *et al.* 2016, Roggeman *et al.* in OHNC, Salillas and Semenza in OHNC and Nieder 2016. The crucial finding of 'number neurons' was reported in papers including Nieder and Miller 2003, Nieder *et al.* 2004 (monkeys; see also Nieder 2013), Nieder *et al.* 2006, Ditz and Nieder 2015 (crows) and Kutter *et al.* 2018 (humans); see further Dehaene *et al.* in HMC, Piazza *et al.* 2004, Nieder and Dehaene 2009, Hyde 2011, Anobile *et al.* 2016, Gallistel 2017, Burr *et al.* 2017, Amalric and Dehaene 2017.

Neural network simulations are discussed (for instance) in Verguts and Fias 2004, Zorzi *et al.* in HMC, Stoianov and Zorzi 2012, Miller 2013, Hannagan *et al.* 2017 and Zorzi and Testolin 2017.

The evolution of number sense is discussed in Beller and Bender 2008, Hubbard *et al.* 2008, Nieder 2018, and Brannon and Park in OHNC.

Subitising is discussed in several of the above reviews; also in Clements 2019. An 'impasse in the literature' is from Agrillo in OHNC, p. 226.

2 *Counting before writing*

Dantzig 1930 should be mentioned as the source underlying much general writing about the history of number, including its origins in human prehistory, though much if not all of its contents are now superseded. Boyer 1944 and Scriba and Ellis 1969 cover some of the same ground; a good recent account of the foundations of number is OHHM chapter 6.1 by Chrisomalis.

Blombos: counting with beads
The Blombos beads are described in D'Errico *et al.* 2005; see also d'Errico *et al.* 2015. Other artefacts from that site are discussed for instance in Henshilwood *et al.* 2001, d'Errico *et al.* 2001, Vanhaeren *et al.* 2013. Other comparable artefacts are discussed in Vanhaeren *et al.* 2006, Bouzouggar *et al.* 2007, d'Errico *et al.* 2008, d'Errico *et al.* 2009; see also the review Álvarez-Fernández 2019. They have become a regular point of reference for wider discussions of embodied or enacted mathematical cognition such as Damerow 2007, Carey 2009 (chapters 4 and 7), Núñez 2009, De Cruz *et al.* 2010, Malafouris 2010, Overmann *et al.* 2011, Coolidge and Overmann 2012, Overmann 2013, 2016c, 2017a, 2017b, 2019, d'Errico *et al.* 2017 and Wynn *et al.* 2017 (related discussions of experimental evidence are in Zhang and Wang 2005 and Lindemann *et al.* 2007), and for more general discussions about the evolution of ('modern') cognition such as Balme and Morse 2006, Kuhn and Stiner 2007, Henshilwood 2007 (and other essays in Mellars *et al.* 2007), Botha 2008, 2016 (chapter 3), Bednarik 2008 and Jeffares 2010. (I have also used accounts that preceded this particular discovery, including Klein 2000, McBrearty and Brooks 2000, d'Errico 2003 and d'Errico *et al.* 2003). For the more general conceptual frameworks involved here see Kirsh 1996, Dartnall 2005, Rowlands 2010, Barrett 2011 and fundamentally Clark and Chalmers 1998, Shapiro 2019 and Malafouris 2013. The

phrase 'wild number line' more usually appears as 'feral number line' in the literature, e.g. in Overmann *et al.* 2011, p. 143; Wynn *et al.* 2013, p. 128.

Lake Rutanzige to Laussel: Counting with tallies
Relevant studies of incised bones are for instance Jones 1873, Davis 1974, d'Errico and Cacho 1994, d'Errico 1995, d'Errico *et al.* 2001, Coinman 1996, Reese 2002, Majkić *et al.* 2018; see also Robinson 1993, Cain 2006, d'Errico and Cacho 1994. For the artefacts from Blombos see above, 'counting with beads', with Henshilwood *et al.* 2009; also Bullington and Leigh 2002 proposing an interpretation of the ochres as tally plaques.

For the Ishango bone see de Heinzelin 1957, 1962, Brooks and Smith 1987, Pletser and Huylebrouck 1999 and Huylebrouck 2008, 2019.

For the woman of Laussel, see Duhard 1988, 1989 in particular; and on images of women in this art more generally, often with specific reference to the Laussel carvings Duhard 1990, 1991, 1993, 1994. 'Unfortunately, the deeper significance of this art will probably remain unknown to us' is from http://www.musee-aquitaine-bordeaux.fr/en/laussel-venus.

Cosquer: Counting by hand
On the date of human dispersal outside Africa, Groucutt *et al.* 2015 is a helpful review.

The role of gesture in the origin and/or grounding of number concepts is addressed in, for example, Fischer 2012, Carey 2009, Gibson 2017 and Malafouris 2010.

The Cosquer cave is described in Clottes *et al.* 1992a, 1992b, Clottes and Courtin 1994, 1996, Collina-Girard 1995 and Sartoretto *et al.* 1995 (with details of its submergence). The second campaign of study is surveyed in Clottes *et al.* 2005a (on which I rely most heavily), 2005b.

Other hand stencils are described in Leroi-Gourhan 1967,

Lorblanchet 1980, Groenen 1988, Barrière and Suères 1993, Delluc and Delluc 1993, Garcia and Duday 1993, Pettitt *et al.* 2015 and García-Diez *et al.* 2015. Interpretations of the hand stencils include Walsh 1979, Groenen 1990, 2011, Gilligan 2010, Cohen 2012, Culley 2016 and McCauley *et al.* 2018; and specifically on the possibility of a counting function Rouillon 2006 and Overmann 2014. 'Uninvited guests' is from Houston 2004, p. 223.

Counting words

On language origins I have used the survey collections Botha and Knight 2009a and 2009b; on the utility of mass comparison and other attempts at deep reconstruction beyond the orthodox language families I follow Campbell and Poser 2008.

Important works on language and number are Hurford 1987 and Wiese 2003, the latter proposing a purely linguistic origin for counting; see also Gelman and Butterworth 2005, Wiese 2007, Chrisomalis 2010a and Wynn *et al.* 2013.

The presently observable universals or near-universals of number words are documented in Greenberg 1978, Corbett 1978, Gvozdanovic 1999, Hammarström 2008a, 2008b, Donohue 2008, WALS s.v. 'Numeral bases' (by Comrie), Comrie 2020 and Epps *et al.* 2012; on ordinals, distributives, etc. see WALS s.v. 'Ordinal Numerals' (by Stolz and Veselinova), 'Distributive Numerals' and 'Numeral Classifiers' (both by Gil). On certain rarities and their implications for these universals see Plank 2009, Hammarström 2010.

The analogy of a bush rather than a line or ladder is from Bahn 1988, p.65.

Interlude: the numbers

A helpful way into questions about what numbers are is provided by the articles in the *Stanford Encyclopedia of Philosophy* (https://

plato.stanford.edu) under 'Philosophy of mathematics', 'fiction-alism', 'formalism', 'intuitionism', 'nominalism', 'structuralism' and 'Platonism'. I have used from the large recent literature on this subject Badiou 2008, De Cruz and De Smedt 2010, De Cruz 2016, Giaquinto 2017 and Overmann 2016b. Allusion is also made to Benacerraf 1973.

3 Counting with words and symbols in the Fertile Crescent

Sumer: Counting symbols

Here I essentially follow Nissen *et al.* 1993 and Robson 2008; see also Chrisomalis 2010b (chapter 7), 2014. My account of the later development of mathematical writing is particularly indebted to Robson 2008 and to Englund 2004; see also the other chapters in Hudson and Wunsch 2004, Katz 2007 (chapter 2 by Robson) and Valério and Ferrara 2020.

Further information on the genesis of writing in this region is taken from Lieberman 1980, Walker 1987, WWS section 3, pp. 37–56 (by Michalowski and Cooper) and Overmann 2016a. I have followed what I take to be the balance of current opinion; for a different interpretation of evidence and timescales see Schmandt-Besserat 1992 and her several other publications. 'People must have realized . . .' is from Van De Mieroop's chapter in Nissen *et al.* 1993, p. 61.

Wider information on the history of the region is in Kuhrt 1995, Van De Mieroop 2004 and CWP chapters 3.7 (Oates), 3.8 (Oates); for its languages see CEWAL (chapters 2 by Michaelowski and 8 by Huehnergard and Woods), and on Akkadian also Goldenberg 2013.

Tiglath-Pileser I: Counting plunder

The fundamental source for this chapter is De Odorico 1995 (see p. 125 for the city of Katmuhu), supplemented by Robson 2008 chapter 5: 'approximation tended to be used . . .' is at p. 69, 'Typically a clean-shaven scribe' at p. 141. Further information on the period and region is from Kuhrt 1995 (see pp. 81–108, 348–65, 473–546), Van De Mieroop 2004 (pp. 89–98, 169–74, 216–52), and also CWP chapters 3.8 (Oates; 'as though it were a footstool' is on p. 1503) and 10 (Özyar), pp. 1545–7. On Tiglath-Pileser I in particular, see Grayson *et al.* 1987, pp. 5–84; 'I captured in battle their king' is from p. 7, 'I gained control over lands' from p. 6.

Details on the Assyrian language are from CEWAL (chapter 8 by Heuhnergard and Woods) and Goldenberg 2013, and on its number words from Goetze 1946 as well as Huehnergard 2011 (pp. 235–8) and pp. 10–11, 30–31, 110 in Goldenberg 2013. Its script is described in WWS pp. 37–56 (by Cooper).

Teianti: Counting coins

For the history of Egypt in this period I rely mainly on Bagnall 1993, CWP chapters 1.16 (Hendrickx) and 1.17 (Ikram), and Erskine 2003 chapters 7 (Thompson) and 15 (Rowlandson); see also Rowlandson 1998. Language and writing are surveyed respectively in CEWAL (chapter 7 by Loprieno) and WWS section 4, pp. 73–83 (by Ritner) (see also Millet 1990 and Dreyer 1992); number and mathematics in Katz 2007 (chapter 1 by Imhausen), Imhausen 2016 ('woe to your limbs' is on p. 144), and Chrisomalis 2010b (chapter 2).

The document discussed here is reproduced, transcribed and translated in Glanville 1939, pp. 43–5 (the quotation is from page 43); discussions of the archive are in Pestman 1989 (pp. 14–23) and at greater length in Depauw 2000. See also https://www.trismegistos.org, record 43769, with a different interpretation of the date.

4 *Counter culture from Athens to the Atlantic*

Philokleon: Counting votes

The general line of thought in this section depends heavily on Netz 2002, whence the phrase 'counter culture' (for a broadly similar take on Roman – specifically Pompeiian – materials, see Bailey 2013). Further discussions of ancient Greek numeracy are in OHHM (chapter 2.1 by Asper) and Cuomo 2001, 2012, 2019 (see also van Berkel 2017 and Sing *et al.* 2021).

Greek number words are discussed in Keyser 2015; see further for the Proto-Indo-European counting words Szemerency 1960, Gvozdanović 1992, Mallory and Adams 2006, pp. 61–2 and Beekes and de Vaan 2011 (chapter 16). Greek rhetorical uses of number are surveyed in Hawke 2008; see also Rubincam 1991, 2001, 2003. The quotation from Demosthenes (27.9–11) appears in Schaps 2004, pp. 18–19.

The Greek number notations are discussed in Tod 1911–12, 1913, 1926–7, 1936–7, 1950, 1954 (all reprinted in Tod 1979) which most subsequent discussions take as a starting point. Recent contributions include Verdan 2017 and Keyser 2018; for the Roman notation Keyser 1988 is also important. Crucially, I follow Chrisomalis 2003 on the relationship of the Greek alphabetic numerals to Egyptian models; see further Chrisomalis 2010b (pp. 140–43).

Athenian courts are discussed in Boegehold 1995, and the exchange of counters therein in Boegehold 1963; the *kleroterion* is also detailed in Bishop 1970, Staveley 1972 and Hansen 1991. Many of the details of court procedure presented here for Philokleon's day may be found in Aristotle's *Athenian Constitution*; his name is from Aristophanes, *Wasps*. Other voting practices such as those at the Assembly are detailed in Kroll 1972, Staveley 1972, Lang 1990 and Hansen 1991.

From the immense literature on Greek coins I have consulted Price 1983, Howgego 1995 (chapter 1), von Reden 1997 and Schaps

2004; see also Haselgrove and Haselgrove and Krmnicek 2012. The lines about 'figs, marshals of the court . . .' appear in Schaps 2004, p. 113, quoting Eubulus *via* Athenaeus XIV 640b–c. On Athenian and wider Greek accounting I have used the essays by Davies, Harris, Kallet-Marx and Langdon in Osborne and Hornblower 1994, and also Cuomo 2013.

Marcus Aurelius: Counting years

The anecdote about Marcus Aurelius is from Cassius Dio, *Roman History* 71.32, translated by Earnest Cary. Bede's line 'very useful and easy' is translated in Wallis 1999, p. 9. The Latin system is discussed in Froehner 1884, Friedlein 1869, Bechtel 1909, Richardson 1916, Cordoliani 1942, 1961, Hilton Turner 1951 (p. 69), Marrou 1958, Alföldi-Rosenbaum 1971, Williams and Williams 1995, Minaud 2006, Weddell 2012 and BNP s.v. 'digitorum computatus'; Bede's account in particular in Jones 1937, 1943 (pp. 329–31) and Cordoliani 1948. On Greek appearances see Schöll 1830 (pp. 345–7), Liddell and Scott 1996 s.v. 'πεμπάζω', Marrou 1958 and Arias and Hirmer 1960, pp. 41–2, 54–5; on Rabanus Maurus see Tannery 1886 and Stevens 1968. For the system in Arabic contexts see Rödiger 1846, Palmer 1869, Lemoine 1932, Welborn 1932, Saidan 1978, pp. 349–50.

'Give me a thousand kisses' is from Catullus 5, 7–8, on which see Minaud 2006; for a counting-board interpretation of the passage see Levy 1941; also (and for the 'impudicus' in another context) Turner 1951.

Blanche of Castile: Counting with silver

Still definitive on the counting board in general, and on *jetons*, is Barnard 1916 (see also Jack 1967). I have also used Sugden 1981, Schärlig 2006 and as well as OCD and BNP *s.v.* 'abacus'. Greek counting boards are discussed in the latter as well as in Liddell and Scott 1996 *s.v.* 'ἄβαξ', and in Schärlig 2001a, 2001b. Controversy about the identity of certain claimed Greek abaci may be seen in

Lang 1957, 1964, 1968 and Pritchett 1965, 1968. 'More a state of mind than an artefact' is from Netz 2002, p. 327.

For the possibility of Mesopotamian counting boards see Woods and Feliu 2017. Roman counting boards are discussed in Levy 1941, Wyatt 1964 and Fellman 1983 as well as in Schärlig 2006; Roman accounting in Scheidel 1996, Cuomo 2011 and Bailey 2013.

For the medieval counting board, Barnard 1916 may be supplemented by Haskins 1912, Baxter 1989, Schärlig 2003 and Periton 2015. I have also used information from the *Oxford English Dictionary* s.vv. 'chess' and 'exchequer'. My account of Köbel 1514 is based directly on that book itself. Chaucer's 'counting-bord' appears in 'The Shipman's Tale', line 83; the Shakespearean insult 'counter-caster' in *Othello* 1.1.31.

More specifically on jetons, I have also used Turckheim-Pey 1997, Barnard 1920, Smith 1921, Freeman 1946, McLellan 2020 and Sarmant and Ploton-Nicollet 2010–.

On Blanche of Castile I have relied mainly on Grant 2016. 'Remarkably adept at ensuring the people did what she wanted' is from p. 11, 'She reacted courageously to challenge' p. 104, 'fashion and charm' p. 319. Her accounts are also discussed in Bougenot 1889.

Number symbols from India

On the name of this set of symbols see Chrisomalis 2010b, p. 1 note 1; I have tried within reason to call the Brahmi/Indian/Toledan/Arabic numerals what the people in my stories called them at the time.

Bhaskara II: Brahmi numerals

For the wider context of writing in South Asia, see WWS section 30 (by Salomon) and CEWAL (chapter 26 by Jamison). On the Arabic numerals in particular, Smith and Karpinski 1911 and

Ginsburg 1917 are still valuable; Chrisomalis 2010b (chapter 6) is my main source.

For the context of mathematics in India I have used Katz 2007 (chapter 4 by Plofker) and Plofker 2009. 'You, radiant Agni' is quoted in Plofker 2009, p. 13, from *Rg-veda* 2.1.8, 'Hail to a hundred' at p. 14, from *Yajur-veda* 7.2.20; for 'the universe is created and destroyed' see p. 53. The *Lilavati* is discussed on pp. 182–91 and elsewhere; see also Plofker 2009b. 'Oh Lilavati, intelligent girl', 'Tell me, quick-eyed girl' and 'There is a hole' are from Katz 2007 (chapter 4 by Plofker), pp. 449, 454 and 459; 'A traveler on a pilgrimage' from Plofker 2009, p. 185.

Counting words in Sanskrit and its descendants are detailed in CEWAL (chapter 26 by Jamison).

Ibn Mun'im: Dust numerals

On the period and location I have used information from Bennison 2016, Abun-Nasr 1987 and Fleet *et al.*, s.v. 'Almohads'; on the sciences in the Arabic-speaking world from Katz 2016 (chapter 3 by Berggren), Brentjes in OHHM, 2018. A system 'which calls for no materials' is from Kunitzsch 2003, p. 5, from Muhammad ibn Yahya al-Sali, *Adab al-kuttab*; 'misbehaved' people from Katz 2016 (chapter 3 by Berggren), p. 532.

Number symbols are discussed in EHAS s.v. 'Numeration and arithmetic' (by Saidan), Kunitzsch 2003 and Burnett 2002a, 2005, 2006; see also the classic paper Gandz 1931. 'Subtle discoveries' is from Burnett 2006, 15; 'the Indians have a most subtle understanding' from p. 17 with note 4 (translation modified).

Ibn Mun'im's combinatorics is discussed in Djebbar 1985, with a French translation of the text. See also Djebbar n.d., EHAS s.v. 'Numeration and arithmetic' (by Saidan), Katz 2007 (chapter 5 by Berggren), pp. 660–5, and Katz 2016 (chapter 3 by Berggren), pp. 427–46.

Hugo of Lerchenfeld: Toledan numerals

On Gerbert's abacus' see Evans 1977a, 1977b, 1979. The later trajectory of Arabic numerals in the Latin world is discussed in Lemay 1982, Burnett 1997 (pp. 48–53 with figures 23–5), EHAS s.v. 'The influence of Arabic mathematics in the medieval West' (by Allard), Berggren 2002, Kunitzsch 2005, Burnett 2002b, Chrisomalis 2010b (chapter 6), Weddell 2015 and Burnett 2020; see also Katz 2016 (chapter 1 by Folkerts and Hughes). For the Hispanic contribution see Lemay 1977; on Greek examples Allard 1977 and Wilson 1981, and on transmission through Antioch Burnett 2000 (pp. 65–66), 2003.

On the practices of calculation with the numerals, see also Allard 1990 and Folkerts 1970 (chapters 9–12), 2001.

The specific case I discuss appears in Arrighi 1968 and Nothaft 2014 (cf. Herreman 2001). Further information on Hugo of Lerchenfeld and the Regensburg annals is from von Fichtenau 1937 (see esp. pp. 321–4) and *Geschichtsquellen des deutschen Mittelalters* (http://www.geschichtsquellen.de) s.vv. 'Annales Ratisponenses', 'Wolfgerus monachus Prufeningensis'.

Material on the later Latin spread of the numerals is from Mercier 1987, Schärlig 2010, Crossley 2013 and Danna 2019. 'Argus, the noble counter' is from Chaucer, *The Book of the Duchess*, line 435. Myths about resistance to the numerals are addressed in Nothaft 2020.

The account keeper: Counting on paper

On Maes see Suchtelen *et al.* 2019; *The Account Keeper* is discussed at pp. 92–5 and also in Rathbone 1951. See also Watkins *et al.* 1984, pp. 218–19 (by Robinson), Robinson 1996 and Eikemeier 1984, and the description at https://www.slam.org/collection/objects/38142/.

The classic studies of ciphering books are Ellerton and Clements

2012, 2014. 'Divide twelve pounds' and the subsequent two quotations are from an American cyphering book of the 1770s described in the latter; see pp. 67, 73, 77. The genre is also studied in Denniss 2012, and in Wardhaugh 2012 (chapter 3).

Caroline Molesworth: Counting the weather

Numerical tabulation is surveyed in Campbell-Kelly *et al.* 2003. Important discussions of the new ways of counting and quantification in the period are Otis 2013, Porter 1995 and Poovey 1998, as well as for the American case Cohen 1982.

The major discussion of weather diaries is Golinski 2007. Specific examples are discussed in Golinski 2001, Brázdil *et al.* 2008, Lee *et al.* 2010, Lorrey 2012, Zhang 2013, Domínguez-Castro *et al.* 2015, Lorrey *et al.* 2016, Thornton *et al.* 2018, Sanderson 2018, Silva 2020, O'Connor *et al.* 2021–2.

The secondary literature on Molesworth is limited to Ormerod 1880. 'Brusqueness and originality' is from p. viii, 'very kind to the poor' p. viii, and 'Miss Molesworth's labours will not have been . . . useless' pp. xxi–xxii. Her journals themselves are online at https://digital.nmla.metoffice.gov.uk/IO_3246bc57-dc56-4ac9-a372-9e5e439a3768/.

Interlude: Number symbols

Menninger 1969, Flegg 1983, Guitel 1975 and (compendiously) Ifrah 1998 discuss the evidence for number symbols; Chrisomalis 2014 is a helpful discussion of what we do not know; the sketch of a classification in this section relies on Chrisomalis 2010b (pp. 9–14). 'There is no ideal numerical notation system' is from Chrisomalis 2010b, p. 19.; 'we do not stand at the end point' from p. 434.

6 Machines that count: Around East Asia

Hong Gongshou: Counting with rods

Background on China's history is from Mote and Twitchett 2008; on its languages CEWAL (chapter 41 by Peyraube) and writing systems WWS section 14 (by Boltz). Numeral notations are discussed in Chrisomalis 2010b (chapter 8); see also Chemla and Ma 2015 and Chemla 2019.

On Ming-period government and economy I have used Mote and Twitchett 2008 chapter 1 (Hucker) and Ma and von Glahn 2022 chapter 9 (Lamouroux and von Glahn); on taxation in the period Huang 1974 and Mote and Twitchett 2008 chapter 2 (Huang), and on the census registers in particular Rhee 2005 and Zhang 2008. The source quoted here is from Rhee 2005; 'the relocation of a particular business tax station' and 'as late as 1578 the imperial university' are from Huang 1974, pp. 176 and 5 respectively. 'A labor service payment of 0.0147445814487 taels of silver' and 'provided a paradise for lower-echelon tax collectors' are both from Mote and Twitchett 2008 chapter 2 (Huang), p. 149.

Discussions of the counting rods and rod numerals are found in Volkov 1994, 2002, 2018; the classic Chinese mathematical texts are also discussed in Libbrecht 1973 (p. 447 for 'Farmer A makes over 407 mou'), Martzloff 1997, OHHM chapter 7.1 (Cullen), Katz 2007 (chapter 3 by Dauben), Dauben 2019 and Yiwen 2020. 'Units are vertical, tens are horizontal' is quoted in Chrisomalis 2010b, p. 264; 'could move his counting-rods as if they were flying' is from Needham and Wang 1959, p. 72; 'the method is complicated' in Katz 2007 (chapter 3 by Dauben), p. 376.

Kiyoshi Matsuzaki (and Thomas Wood): Counting with beads

Primary sources for the contest are Anon 1946 and Kojima 1954; the step backwards for 'the machine age' and '"civilization" had "tottered"' are both quoted on p. 12. Further background information is taken from Duus 2008 esp. chapters 4 (Fukui), 7 (Coox), 10 (Kōsai and Goble); Chrisomalis 2020 (p. 74) mentions the earlier tradition of such contests. Another important primary source for *soroban* use in the mid-twentieth century is Kojima 1963; useful histories include Pullan 1968 and Schärlig 2006. 'Children go there after school' is from Stigler *et al.* 1986, p. 454.

'The character "Suan"' is from Dauben 1998, p. 1348 note 12. Cognitive aspects of the *suanpan* and *soroban* are discussed in Hatano *et al.* 1977, Stigler *et al.* 1986, Hatta *et al.* 1989, Hishitani 1990, Miller and Stigler 1991, Yujie *et al.* 2020 and Wang 2020. Related but more general issues in visual processing are reported in Alvarez and Cavanagh 2005, Halberda *et al.* 2006 and Feigenson 2008. More specifically on the mental abacus, see Stigler *et al.* 1982, Stigler 1982, 1984, Hatano and Osawa 1983, Hatano *et al.* 1987, Frank and Barner 2011, Kim 2016, Barner *et al.* 2016, Cho and So 2018, Brooks *et al.* 2018 and Cheng *et al.* 2021.

Hermann Hollerith and Kawaguchi Ichitaro: Counting machines
Kawaguchi's Electric Tabulation Machine is pictured and described at https://museum.ipsj.or.jp/en/computer/dawn/0056.html and https://museum.ipsj.or.jp/en/heritage/denkishukeiki.html. On the Japanese census see Matsuda 1981.

On the 1890 US census, important primary sources are Walker 1888, Wright 1889, Anon 1890 ('The blanks which had been filled up' is on p. 132) and Porter 1891, 1894. Historical reflections from soon after that census include Hunt 1899, Wright 1900 (pp. 69–79), SHCA 1903 (see pp. 173–4) and Wilcox 1914. 'A very tidy and airy machineshop' and 'your tympanums all tingle' are quoted in Austrian 1982, pp. 60–1, respectively from the *Washington Star* of 26 June 1890 and the *Saginaw (Michigan) Weekly* of 23 October 1890.

'The unsettled area' of the American continent is from Turner 1986, p. 1. Modern histories of the US census I have used include

Anderson 1988, USCB 2002, Schor 2017 and Ruggles and Magnuson 2020. The loss of the 1890 records is discussed in O'Mahony 1991, Blake 1996 and Dorman 2008; 'might every one of them be burned up' is from Martin 1891, p. 525.

Hollerith's own descriptions of his inventions appear in Hollerith 1889a, 1889b, 1894 among many other sources. Other (relatively) technical discussions from the period are Martin 1891, Newcomb 1913 (adoption by 'nearly every civilized country' on p. 82) and Felt 1916. From the voluminous more recent secondary literature I have used Truesdell 1965, Bashe 1986 (with one of the fullest descriptions of how the system actually worked), Reid-Green 1989, Campbell-Kelly 1989, Norberg 1990, Kistermann and Reston 1991, USCB 1991 and Campbell-Kelly *et al.* 2014. Additionally on Hollerith himself I have used Austrian 1982 and *American National Biography* (https://www.anb.org).

On the later development of the punched card I have consulted Lubar 2004 and Heide 2008, 2009.

Sia Yoon: Counting likes
On the history of computers I have used Akera and Nebeker 2002, Haigh and Ceruzzi 2021, O'Regan 2008 and Campbell-Kelly *et al.* 2014; on the digital revolution Hobart and Schiffman 1998, Deuze 2006, Gere 2008, Clarke 2012 and George 2020. 'Consensual hallucination' is from *Neuromancer* (William Gibson, 1984), p. 5.

On Korean history I use Connor 2002, Kim 2017 (especially Zur's chapter) and Seth 2020; on technology therein Choi 2017. On K-pop and Korean Manga, Lie 2014, Pasfield-Neofitou *et al.* 2016 and Jin 2016, and on the webtoon phenomenon Jin 2015 ('on a bustling subway ride to work' is from p. 193), Kim and Yu 2019, Yecies *et al.* 2020, Shim *et al.* 2020 and Cho 2021. Sia's story is online at https://www.webtoons.com/en/drama/my-dud-to-stud-boyfriend/list?title_no=4353.

7 Counting words and more in the Pacific world

Ayangkidarrba: Counting eggs

On the Australian archaeology I have used CWP chapters 1.35 (White) and 1.37 (Bird), and Cochrane and Hunt 2018 (chapters by Cochrane and Hunt, and O'Connor). An important recent publication on the earliest archaeology in the continent is Clarkson *et al.* 2017, with accompanying comments and reply (Clarkson *et al.* 2018).

A key anthropological source for this chapter is Keen 2004. On the languages of the continent it is important to be aware on the one hand of Dixon 1980, 2002, and on the other of such accounts as Bowern and Koch 2004, Campbell and Poser 2008 (pp. 318–25), and CWP chapter 1.42 (Heggarty and Renfrew). Additional information is taken from Hale 1975, Evans 2003, Cunningham *et al.* 2006 (chapters 5 and 12), Miyaoka *et al.* 2007 (chapter 12 by Walsh) and Leitner and Malcolm 2007. On counting in Australian languages I use Harris 1982, 1987, Sayers 1982, McRoberts 1990, Bowern and Zentz 2012, Epps *et al.* 2012, Treacy *et al.* 2015 and Zhou and Bowern 2015.

Specifically on Groote Eylandt and the Anindilyakwa I have used Worsley 1954, Turner 1974 ('would awake at sunrise': p. 165), and for the contact with Macassan people May *et al.* 2009. On their language, Leeding 1989 and van Egmond and Baker 2020. On their counting Stokes 1982, which I follow, describing the situation of number words and counting documented there (including rejecting a Macassan source for the lexemes and the base 5 system). 'The hand is held loosely' and 'there is a story on Groote Eylandt' are from Stokes 1982, pp. 43 and 39. Butterworth *et al.* 2008 gives a different account of the number words in this language and states that they are not (now) taught to children, presumably reflecting recent change.

Oksapmin: body count

For the context of this chapter in the ethnography and history of Papua New Guinea I have consulted Brongersma and Venema 1962 and Gammage 1998, and on the languages of the region the remarks in CWP chapter 1.42 (Heggarty and Renfrew), Dixon 2002 and Campbell and Poser 2008, as well as (particularly) Lean 1992 and Owens *et al.* 2018. On the Oksapmin in particular I have used Perey 1973 (p. 27 for 'a raft crossing described as "very risky"'). On the Oksapmin language, see also Loughnane and Fedden 2011.

On counting in New Guinea, the classic and monumental studies are Lean 1992 and Owens 2001, synthesised in Owens *et al.* 2018. I have also referred to Laycock 1975, De Vries 2014 and Dwyer and Minnegal 2016, and to the classic Mimica 1988. On the cognitive and educational side, Matang 2005 and Matang and Owens 2014 are valuable.

Oksapmin counting in particular has been exhaustively studied in Saxe 1981, 1982, 1999, Saxe and Esmonde 2005 and the synthesis Saxe 2012; see also https://culturecognition.com. '*Tipana, tipnarip, bumrip, hadrip, hatatah*' are quoted in the form given at Saxe 2012, p. 271; see p. 45 for a slightly different transcription. 'All types of useless and unnecessary items' is quoted in Saxe 2012, p. 51. The material about change in the Oksapmin number system is derived primarily from Saxe 1982, 2012 and Saxe and Esmonde 2005 (on *fu* in particular). 'In 1960, when Australian shillings and pounds were used' is from Saxe and Esmonde 2005, p. 209–10; 'some people whose participation in the money economy' from Saxe 2012, p. 30.

Tonga: Counting leaves

On the prehistory of the Pacific islands I have used the classic account Terrell 1986 as well as Kirch 2000 (whence the phrase 'path of the winds'), Oppenheimer 2004, CWP chapter 1.31 (Tanudirjo) and Hunt and Cochrane 2018 (articles by Cochrane and Hunt, Terrell, Rieth and Cochran, Burley and Addison, the last having particular relevance to Tonga). On Tongan culture in the historical period, important documents are Mariner 1818, Gifford 1929, Bott and Tavi 1982; the more recent historical writing I have consulted includes Rutherford 1977, Petersen 2000, Campbell 2001 and Evans 2001.

On the Austronesian language family I have used Lynch 2018 as well as CWP chapter 1.33 (Heggarty and Renfrew). Number words in Oceanic languages are discussed in Harrison and Jackson 1984, Bender *et al.* 2006, Bender 2013 and Bender and Beller 2006a, 2006b, 2014, 2018, 2021; I rely particularly on their 2006b for the interpretation of the complex history of Polynesian and Micronesian numeral classifiers and counting systems. Specifically on Tongan number words I use Bender and Beller 2007; 'the kie leaves are cut while still green' and the two subsequent quotations are from Bender and Beller 2007, 230, 'one such presentation' from p. 231 and 'some do not apply or even remember the traditional systems' from p. 229. 'On Lamotrek and Fais' is from Bender and Beller 2006b, p. 400; 'breadfruit, pandanus leaves' from p. 385.

8 *Panorama: Counting in the Americas*

On the archaeology of the Americas, I have used CWP chapter 2.14 (Collins) together with Meltzer 2021. General information about American languages is from Mithun 1999, CWP chapter 2.36 (Heggarty and Renfrew) and Miyaoka *et al.* 2007; that on American numeracies from Chrisomalis 2010b, chapter 9, Closs 1986 and DeCesare 1999.

Yup'ik: Counting games
On the Yup'ik archaeology see Shaw 1998 and Masson-MacLean *et al.* 2020; also on the region, CWP chapter 2.15 (Anderson). The ongoing excavation of Nunalleq is documented at nunalleq.word-press.com, where the bundle of tally sticks is discussed (nunalleq.

wordpress.com/2018/07/23/artfact-of-the-day-20-july/); other sources used are Knecht and Jones 2019, Mossolova and Knecht 2019 (on masks) and Sloan 2021. 'Whittled bits of wood, sharpened stakes' and 'protruding from the dark soils' are from Knecht and Jones 2019, 28.

Historical documentation of ways of life in the region appears in Nelson 1899; the descriptions of games including the three quotations are from p. 332. See also Funk 2010 on the 'Bow and Arrow War'; ethnographic sources consulted include Fienup-Riordan 1990, Barker and Barker 1993, www.yupikscience.org, Williams 2009, Fienup-Riordan and Rearden 2012 and Jolles and Oozeva 2002.

Counting in Yup'ik is discussed particularly in Koo and Bartman 1980; see also (including on related languages) Baillargeon *et al.* 1977 ('"How long was it", I asked': p. 126), Closs 1986 (pp. 129–80 by Denny), Lipka 1994 and Corbett and Mithun 1996. The tally tattoo mentioned is from Carrillo 2014, p. 65; several other tally objects can be seen online at the National Museum of the American Indian, Smithsonian Institution (americanindian.si.edu).

Pomo: Counting costs

Historical and ethnographic sources on the Pomo include Powers 1877, Barrett 1908, Loeb 1926, Stewart 1943, Barrett 1952 and Colson 1974; on the number words in the seven Pomo languages see Dixon and Kroeber 1907 (esp. pp. 676, 685, 686).

Historical sources on shell money in the region include Stearns 1869, 1877; Pomo bead making is discussed in Loeb 1926 including at pp. 176–8, 191–5, 229–30, and in Hudson 1897 and Gifford 1926 (pp. 329, 377–8, 386–7; p. 378 for 'the Southeastern Pomo informant Wokox'). On the economic practices of the region more generally Davis 1961 and Vayda 1967 are useful; 'the host chief divided up' is from Loeb 1926, p. 193 and 'two funerals in one month' from Colson 1974, p. 205. More recent work on the Pomo money and its history includes Heizer 1975, Arnold and Munns 1994, Parker 2010, Burns 2019, Trubitt 2003 and Gamble 2020.

Waxaklahun-Ubah-K'awil: The long count

On the general history of the Maya I have used Sharer and Traxler 2006; on the Mayan languages CWP chapter 2.36 (Heggarty and Renfrew) and CEWAL (chapter 43 by Bricker); and on their scripts WWS section 12 (by Macri). Mathematical notations are also discussed in Closs 1986 (pp. 291–370) and in Chrisomalis 2010b (chapter 9). The question of the Mayan zero is discussed in Justeson 2010 and Blume 2011 in addition to Chrisomalis 2010b, which I follow on the question of whether and in what sense the notation was positional.

The Mayan calendars are discussed in several of the above works, and also in Gillispie *et al.* 2008, s.v. 'Maya Numeration', Aveni 2001, Jones 2005, s.v. 'Calendars: Mesoamerican Calendars', Pharo 2014, Normark 2016 and Milbrath 2017. On Copán in particular I have consulted Fash and Fash 1991/2001, Schele and Mathews 1998 (chapter 4), Wyllys and Fash 2005 and especially Newsome 1991. 'His breath expired in war' is quoted in Blume 2011, p. 51. Stela D is also discussed in Pineda De Carías *et al.* 2017.

Pirahã: Lost count

The original report of the Pirahã's lack of number words is Gordon 2004; 'clusters of nuts matched to the battery line' is from p. 498 and 'I sat across from the participant' p. 497. See Pica *et al.* 2004 for a discussion and Everett 2005 for further experimental work. 'The Pirahã are some of the brightest' is from p. 621; 'could not find anyone' p. 632; 'someone can ask for an entire roll of hard tobacco' and 'in this "trade relationship"' p. 626; and 'I prefer whole animals' p. 625.

Questions about the original observations are raised in Frank *et al.* 2008; see Everett and Madora 2012 for a reply.

Select Bibliography

Abun-Nasr, Jamil M., *A History of the Maghrib in the Islamic Period* (Cambridge, 1987).

Akera, Atsushi and Frederik Nebeker, *From 0 to 1: An Authoritative History of Modern Computing* (Oxford, 2002).

Alföldi-Rosenbaum, Elisabeth, 'The finger calculus in antiquity and in the Middle Ages', *Frühmittelalterliche Studien* 5 (1971), pp. 1–9.

Allard, André, 'Le premier traité byzantin de calcul indien: classement des manuscrits et édition critique du texte', *Revue d'histoire des textes* 7 (1977), pp. 57–107.

Allard, André, 'La formation du vocabulaire latin de l'arithmétique médiévale', in Olga Weijers, ed., *Méthodes et instruments du travail intellectuel au moyen âge: études sur le vocabulaire* (Turnhout, 1990), pp. 137–81.

Alvarez, G. A. and P. Cavanagh, 'Independent resources for attentional tracking in the left and right visual hemifields', *Psychological Science* 16 (2005), pp. 637–43.

Álvarez-Fernández, Esteban, ed., *Special Issue: Personal Ornaments in Early Prehistory: A Review of Shells as Personal Ornamentation during the African Middle Stone Age, PaleoAnthropology* (2019).

Amalric, M. and S. Dehaene, 'Cortical circuits for mathematical knowledge: evidence for a major subdivision within the brain's semantic networks', *Philosophical Transactions B* 373 (2017).

Anderson, Margo J., *The American Census: A Social History* (New Haven, 1988).

Anobile, G., *et al.*, 'Number as a primary perceptual attribute: a review', *Perception* 45 (2016), pp. 5–31.

Anon., 'To think or not to think?', *Time* (25 November 1946), p. 35.

Anon., 'The Census of the United States', *Scientific American* 63 (30 August 1890), p. 132.

Arias, Paolo E. and Max Hirmer, *Tausend Jahre griechische Vasenkunst* (Munich, 1960).

Arnold, Jeanne E. and Ann Munns, 'Independent or attached specialization: the organization of shell bead production in California', *Journal of Field Archaeology* 21 (1994), pp. 473–89.

Arrighi, G., 'La numerazione "arabica" degli Annales Ratisbonenses', *Physis* 10 (1968), pp. 243–57.

Austrian, G., *Herman Hollerith: Forgotten Giant of Information Processing* (New York, 1982).

Aveni, Anthony, 'Time, number, and history in the Maya world', *KronoScope: Journal for the Study of Time* 1 (2001), pp. 29–61.

Badiou, Alain, *Number and Numbers* (Cambridge, 2008).

Bagnall, R., *Egypt in Late Antiquity* (Princeton, 1993).

Bahn, Paul G., *Images of the Ice Age* (Oxford, 1988).

Bailey, Melissa, 'Roman money and numerical practice', *Revue belge de Philologie et d'Histoire* 91 (2013), pp. 153–86.

Baillargeon, R. *et al.*, 'Aspects sémantiques et structuraux de la numération chez les Inuit', *Etudes Inuit* 1 (1977), pp. 93–128.

Balme, J. and K. Morse, 'Shell beads and social behaviour in Pleistocene Australia', *Antiquity* 80 (2006), pp. 799–811.

Barker, J. H. and R. Barker, *Always Getting Ready / Upterrlainarluta: Yup'ik Eskimo Subsistence in Southwest Alaska* (Seattle, 1993).

Barnard, F. P., *The Casting-Counter and the Counting-Board* (Oxford, 1916).

Barnard, F. P., 'Italian jettons', *Numismatic Chronicle* 20 (1920), pp. 216–72.

Barner, David, *et al.*, 'Learning mathematics in a visuospatial format: a randomized, controlled trial of mental abacus instruction', *Child Development* 87 (2016), pp. 1146–58.

Barrett, L., *Beyond the Brain: How Body and Environment Shape Human and Animal Minds* (Princeton, 2011).

Barrett, Samuel Alfred, *The Ethno-geography of the Pomo and Neighboring Indians* (Berkeley, CA, 1908).

Barrett, Samuel Alfred, *Material Aspects of Pomo Culture* (Milwaukee, 1952).

Barrière, C. and M. Suères, 'Les mains de Gargas', *Les Dossiers d'Archéologie* 178 (1993), pp. 46–55.

Bashe, Charles J., *IBM's early computers* (Cambridge, MA, 1986).

Baxter, W. T., 'Early accounting: the tally and checkerboard', *The Accounting Historian's Journal* 16 (1989), pp. 43–83.

Bechtel, Edward A., 'Finger-counting among the Romans in the fourth century', *Classical Philology* 4 (1909), pp. 25–31.

Bednarik, R. G., 'Beads and cognitive evolution', *Time and Mind: The Journal of Archaeology, Consciousness and Culture* 1 (2008), pp. 285–318.

Beekes, R. S. P. and Michiel Arnoud Cor de Vaan, *Comparative Indo-European Linguistics: An Introduction* (2nd edition: Amsterdam, 2011).

Beller, S. and A. Bender, 'The limits of counting: numerical cognition between evolution and culture', *Science* 319 (2008), pp. 213–15.

Benacerraf, Paul, 'Mathematical truth', *Journal of Philosophy* 70 (1973), pp. 661–79.

Bender, A., 'Two accounts of Mangarevan counting . . . and how to evaluate them', *Journal of the Polynesian Society* 122 (2013), pp. 275–87.

Bender, A., *et al.*, 'The cognitive advantages of counting specifically: a representational analysis of verbal numeration systems in Oceanic languages', *Topics in Cognitive Science* 7 (2006), pp. 552–69.

Bender, A. and S. Beller, '"Fanciful" or genuine? Bases and high numerals in Polynesian number systems', *The Journal of the Polynesian Society* 115 (2006a), pp. 7–46.

Bender, A. and S. Beller, 'Numeral classifiers and counting systems in Polynesian and Micronesian languages: common roots and cultural adaptations', *Oceanic Linguistics* 45 (2006b), pp. 380–403.

Bender, A. and S. Beller, 'Counting in Tongan: The traditional number systems and their cognitive implications', *Journal of Cognition and Culture* 7 (2007), pp. 213–39.

Bender, A. and S. Beller, 'Mangarevan invention of binary steps for easier calculation', *Proceedings of the National Academy of Sciences of the United States of America* 111 (2014), pp. 1322–7.

Bender, A. and S. Beller, 'Numeration systems as cultural tools for numerical cognition', in D. B. Berch *et al.*, eds., *Mathematical Cognition and Learning: Language and Culture in Mathematical Cognition* (Cambridge, MA, 2018), pp. 297–320.

Bender, A. and S. Beller, 'Ways of counting in Micronesia', *Historia Mathematica* 56 (2021), pp. 40–72.

Bennison, Amira K., *The Almoravid and Almohad Empires* (Edinburgh, 2016).

Benson-Amram, S., *et al.*, 'Numerical assessment in the wild: insights from social carnivores', *Philosophical Transactions B* 373 (2017).

Berggren, J. Lennart, 'Medieval arithmetic: Arabic texts and European motivations', in John J. Contreni and Santa Casciani, eds., *Word, Image, Number: Communication in the Middle Ages* (Florence, 2002), pp. 351–65.

Bishop, J. D., 'The Cleroterium', *Journal of Hellenic Studies* 90 (1970), pp. 1–14.

Blake, Kellee, '"First in the path of the firemen": The fate of the 1890 population census', *Prologue Magazine* 28 (Spring 1996), parts 1–2.

Blume, Anna, 'Maya concepts of zero', *Proceedings of the American Philosophical Society* 155 (2011), pp. 51–88.

Boegehold, Alan L., 'Toward a study of Athenian voting procedure', *Hesperia* 32 (1963), pp. 366–74.

Boegehold, Alan L., *The Lawcourts at Athens* (Princeton, 1995).

Botha, R., 'Prehistoric shell beads as a window on language evolution', *Language & Communication* 28 (2008), pp. 197–212.

Botha, R., *Language Evolution: The Windows Approach* (Cambridge, 2016).

Botha, R. and Chris Knight, *The Prehistory of Language* (Oxford, 2009a).

Botha, R. and Chris Knight, *The Cradle of Language* (Oxford, 2009b).

Bott, E., with the assistance of Tavi, *Tongan Society at the Time of Captain Cook's Visits: Discussions with Her Majesty Queen Salote Tupou* (Wellington, 1982).

Bougenot, Etienne Symphorien, 'Comptes de dépenses de Blanche de Castille', *Bulletin du Comité des travaux historiques et scientifiques: section d'histoire et de philologie* (1889), pp. 86–91.

Bouzouggar, A. *et al.*, '82,000-year-old shell beads from North Africa and implications for the origins of modern human behavior', *Proceedings of the National Academy of Sciences of the United States of America* 104 (2007), pp. 9964–9.

Bowern, C. and H. Koch, *Australian Languages: Classification and the Comparative Method* (Amsterdam, 2004).

Bowern, Claire and Jason Zentz, 'Diversity in the Numeral Systems of Australian Languages', *Anthropological Linguistics* 54 (2012), pp. 133–60.

Boyer, Carl B., 'Fundamental steps in the development of numeration', *Isis* 35 (1944), pp. 153–65.

Brázdil, R. *et al.*, 'Weather information in the diaries of the Premonstratensian Abbey at Hradisko, in the Czech Republic, 1693–1783', *Weather* 63 (2008), pp. 201–7.

Brentjes, S., *Teaching and Learning the Sciences in Islamicate Societies (800–1700)* (Turnhout, 2018).

Brongersma, L. D. and G. F. Venema, *To the Mountain of the Stars* (London, 1962).

Brooks, Alison S. and Catherine C. Smith, 'Ishango revisited: new age determinations and cultural interpretations', *The African Archaeological Review* 5 (1987), pp. 65–78.

Brooks, N. B., *et al.*, 'The role of gesture in supporting mental representations: The case of mental abacus arithmetic', *Cognitive Science* 42 (2018), pp. 554–75.

Bullington, Jill and Steven R. Leigh, 'Rock art revisited' [letter], *Science* 296 (2002), p. 468.

Burnett, Charles, *The Introduction of Arabic Learning into England* (London, 1997).

Burnett, Charles, 'Antioch as a link between Arabic and Latin culture in the twelfth and thirteenth centuries', in A. Tihon *et al.*, eds, *Occident et Proche-Orient: contacts scientifiques au temps des croisades* (Louvain-la-Neuve, 2000), pp. 1–78.

Burnett, Charles, 'Indian numerals in the Mediterranean Basin in the twelfth century, with special reference to the "Eastern forms"', in Y. Dold-Samplonius *et al.*, eds., *From China to Paris: 2000 years' transmission of mathematical ideas* (Stuttgart, 2002a), pp. 237–88.

Burnett, Charles, 'The abacus at Echternach in ca. 1000 A.D.', *SCIAMVS* 3 (2002b), pp. 91–108.

Burnett, Charles, 'The use of Arabic numerals among the three language cultures of Norman Sicily', *Römisches Jahrbuch der Bibliotheca Hertziana* 35 (2005), pp. 39–48.

Burnett, Charles, 'The semantics of Indian numerals in Arabic, Greek and Latin', *Journal of Indian Philosophy* 34 (2006), pp. 15–30.

Burnett, Charles, 'The Palaeography of Numerals', in Frank T. Coulson and Robert G. Babcock, eds, *The Oxford Handbook of Latin Palaeography* (Oxford, 2020), pp. 25–36.

Burns, Gregory, 'Evolution of Shell Bead Money in Central

California: An Isotopic Approach', unpublished Ph.D. dissertation, University of California, 2019.

Burr, D. C., *et al.*, 'Psychophysical evidence for the number sense', *Philosophical Transactions B* 373 (2017).

Butterworth, B., *et al.*, 'Numerical thought with and without words: evidence from indigenous Australian children', *Proceedings of the National Academy of Sciences* 105 (2008), pp. 13179–84.

Butterworth, B., *et al.*, eds, 'Discussion meeting issue "The origins of numerical abilities"', *Philosophical Transactions B* 373 (2018).

Cain, C. R., 'Implications of the marked artifacts of the Middle Stone Age of Africa', *Current Anthropology* 47 (2006), pp. 675–81.

Campbell, I. C., *Island Kingdom: Tonga Ancient and Modern* (Christchurch, 2001).

Campbell, Lyle and William J. Poser, *Language Classification: History and Method* (Cambridge, 2008).

Campbell-Kelly, Martin, 'Punched-card machinery', in William Aspray, ed., *Computing before Computers* (Ames, 1989), pp. 122–55.

Campbell-Kelly, Martin, *et al.*, *Computer: A History of the Information Machine* (3rd edn: Boulder, CO, 2014).

Campbell-Kelly, Martin, *et al.*, eds, *The History of Mathematical Tables: From Sumer to Spreadsheets et al.* (Oxford, 2003).

Cantlon, J. F. and E. M. Brannon, 'Basic math in monkeys and college students', *PLoS biology* 5 (2007a), e328.

Cantlon, J. F. and E. M. Brannon, 'How much does number matter to a monkey (Macaca mulatta)?', *Journal of Experimental Psychology: Animal Behavior Processes* 33 (2007b), pp. 32–41.

Carey, S., *The Origin of Concepts* (New York, 2009).

Carrillo, Mariah, 'Transformative skin: the ongoing legacy of Inuit and Yupik women's tattoos', unpublished MA dissertation, University of New Mexico, 2014.

Chemla, Karine, 'Different clusters of text from ancient China, different mathematical ontologies', *HAU: Journal of Ethnographic Theory* 9 (2019), pp. 99–112.

Chemla, Karine and B. Ma, 'How do the earliest known mathematical writings highlight the state's management of grains in early imperial China?', *Archive for History of Exact Sciences* 69 (2015), pp. 1–53.

Cheng, Dazhi *et al.*, 'Chinese kindergarteners skilled in mental abacus have advantages in spatial processing and attention', *Cognitive Development* 58 (2021).

Cho, Heekyoung, 'The platformization of culture: webtoon platforms and media ecology in Korea and beyond', *The Journal of Asian Studies* 80 (2021), pp. 73–93.

Cho, Philip S. and Wing Chee So, 'A feel for numbers: the changing role of gesture in manipulating the mental representation of an abacus among children at different skill levels', *Frontiers in Psychology* 9 (2018), Article 1267.

Choi, Hyungsub, 'The Social Construction of Imported Technologies: Reflections on the Social History of Technology in Modern Korea', *Technology and Culture* 58 (2017), pp. 905–20.

Chrisomalis, Stephen, 'The Egyptian origin of the Greek alphabetic numerals', *Antiquity* 77 (2003), pp. 485–96.

Chrisomalis, Stephen, 'Indiscrete infinities: numerical representations and the evolution of language', paper presented at the 40th Annual Meeting of the Michigan Linguistics Society (Flint, MI, 2010a).

Chrisomalis, Stephen, *Numerical Notation: A Comparative History* (Cambridge, 2010b).

Chrisomalis, Stephen, 'Six Unresolved Questions in the Early History of Numeration', paper presented at Signs of Writing: The Cultural, Social, and Linguistic Contexts of the World's First Writing Systems Neubauer Collegium for Culture and Society (Chicago, IL, 2014).

Chrisomalis, Stephen, *Reckonings: Numerals, Cognition, and History* (Cambridge, MA, 2020).

Clark, A. and D. Chalmers, 'The extended mind', *Analysis* 58 (1998), pp. 7–19.

Clarke, Michael, 'The digital revolution', in Robert Campbell *et al.*, eds, *Academic and Professional Publishing* (Oxford, 2012), pp. 79–98.

Clarkson, C., *et al.*, 'Human occupation of northern Australia by 65,000 years ago', *Nature* 547 (2017), pp. 306–25.

Clarkson, C., *et al.*, Reply to comments on Clarkson *et al.* (2017), 'Human occupation of northern Australia by 65,000 years ago', *Australian Archaeology* 84 (2018), pp. 84–89.

Clements, Douglas H., 'Subitizing: The neglected quantifier', in A. Norton and M. W. Alibali, eds, *Constructing Number* (Cham, 2019).

Closs, M., ed., *Native American Mathematics* (Austin, TX, 1986).

Clottes, J. and J. Courtin, *La grotte Cosquer. Peintures et gravures de la caverne engloutie* (Paris, 1994).

Clottes, J. and J. Courtin, *The Cave Beneath the Sea: Paleolithic Images at Cosquer* (New York, 1996).

Clottes, J. *et al.*, 'The Cosquer Cave on Cape Morgiou, Marseilles', *Antiquity* 66 (1992a), pp. 583–98.

Clottes, J. *et al.*, 'La grotte Cosquer datée', *Bulletin de la Société préhistorique française* 89 (1992b), pp. 230–234.

Clottes, J. *et al.*, *Cosquer redécouvert* (Paris, 2005a).

Clottes, Jean *et al.*, 'Nouvelles recherches a la Grotte Cosquer (Marseille)', *Munibe Antropologia-Arkeologia* 57 (2005b), pp. 9–22.

Cochrane, Ethan E. and Terry L. Hunt, *The Oxford Handbook of Prehistoric Oceania* (Oxford, 2018).

Cohen, Patricia Cline, *A Calculating People: The Spread of Numeracy in Early America* (Chicago, 1982).

Cohen, Claudine, 'Symbolique de la main dans l'art pariétal paléolithique', *Académie des beaux-arts* (2012), pp. 61–73.

Coinman, N. R., 'Worked bone in the Levantine Upper Paleolithic: Rare examples from the Wadi al-Hasa, West-Central Jordan', *Paléorient* 22 (1996), pp. 113–21.

Collina-Girard, J., 'La Grotte Cosquer et les sites paléolithiques du littoral marseillais (entre Carry le Rouet et Cassis)', *Méditerranée* 3 (1995), pp. 7–19.

Colson, Elizabeth, *Autobiographies of Three Pomo Women* (Berkeley, CA, 1974).

Comrie, Bernard, 'Revisiting Greenberg's "Generalizations about Numeral Systems" (1978)', *Journal of Universal Language* 21 (2020), pp. 43–84.

Connor, Mary E., *The Koreas: A Global Studies Handbook* (Santa Barbara, CA, 2002).

Coolidge, F. L. and Karenleigh A. Overmann, 'Numerosity, Abstraction, and the Emergence of Symbolic Thinking', *Current Anthropology* 53 (2012), pp. 204–25.

Corbett, Greville G., 'Universals in the syntax of cardinal numerals', *Lingua* 46 (1978), pp. 355–68.

Corbett, Greville G. and Marianne Mithun, 'Associative Forms in a Typology of Number Systems: Evidence from Yup'ik', *Journal of Linguistics* 32 (1996), pp. 1–17.

Cordoliani, A., 'Etudes du comput I: Note sur le manuscrit 7418 de la Bibliotheque Nationale', *Bibliothèque de l'École des chartes* 103 (1942), pp. 61–5.

Cordoliani, A., 'A propos du Chapitre premier du *De Temporum ratione*, de Bede', *Le Moyen Age* 54 (1948), pp. 209–23.

Cordoliani, A., 'Les manuscrit de comput des bibliotheques d'Utrecht', *Scriptorium* 15 (1961), pp. 76–85.

Crossley, John N., 'Old-fashioned versus newfangled: reading and writing numbers, 1200–1500', *Studies in Medieval and Renaissance History* 10 (2013), pp. 79–109.

Crump, T., *The Anthropology of Numbers* (Cambridge, 1990).

Culley, Elisabeth V., 'A Semiotic Approach to the Evolution of Symboling Capacities During the Late Pleistocene with Implications for Claims of "Modernity"', unpublished PhD dissertation, Arizona State University, 2016.

Cunningham, D. *et al.*, *Language Diversity in the Pacific: Endangerment and Survival* (Clevedon, 2006).

Cuomo, Serafina, *Ancient Mathematics* (London, 2001).

Cuomo, Serafina, 'All the proconsul's men: Cicero, Verres, and account-keeping', in A. Roselli, ed., *L'insegnamento delle technai nelle culture antiche* (Pisa, 2011), pp. 165–88.

Cuomo, Serafina, 'Exploring ancient Greek and Roman numeracy', *BSHM Bulletin: Journal of the British Society for the History of Mathematics* 27 (2012), pp. 1–12.

Cuomo, Serafina, 'Accounts, numeracy and democracy in classical Athens', in M. Asper, ed., *Writing Science: Medical and Mathematical Authorship in Ancient Greece* (Berlin, 2013), pp. 255–78.

Cuomo, Serafina, 'Mathematical traditions in Ancient Greece and Rome', *HAU Journal of Ethnographic Theory* 9 (2019), pp. 75–85.

d'Errico, F., 'A new model and its implications for the origin of writing: The La Marche antler revisited', *Cambridge Archaeological Journal* 5 (1995), pp. 163–206.

d'Errico, F., 'The invisible frontier: a multiple species model for the origin of behavioural modernity', *Evolutionary Anthropology* 12 (2003), pp. 182–202.

d'Errico, F. and C. Cacho, 'Notation versus decoration in the Upper Paleolithic: A case-study from Tossal de la Roca, Alicante, Spain', *Journal of Archaeological Science* 21 (1994), pp. 185–200.

d'Errico, F. *et al.*, 'An engraved bone fragment from c. 70,000-year-old Middle Stone Age levels at Blombos Cave, South Africa: implications for the origin of symbolism and language', *Antiquity* 75 (2001), pp. 309–18.

d'Errico, F. *et al.* 'Archaeological evidence for the emergence of language, symbolism, and music: an alternative multidisciplinary perspective', *Journal of World Prehistory* 17 (2003), pp. 1–70.

d'Errico, F. *et al.*, '*Nassarius kraussianus* shell beads from Blombos

Cave: Evidence for symbolic behaviour in the Middle Stone Age', *Journal of Human Evolution* 48 (2005), pp. 3–24.

d'Errico, F. *et al.*, 'Possible shell beads from the Middle Stone Age layers of Sibudu Cave, South Africa', *Journal of Archaeological Science* 35 (2008), pp. 2675–85.

d'Errico, F. *et al.*, 'Additional evidence on the use of personal ornaments in the Middle Paleolithic of North Africa', *Proceedings of the National Academy of Sciences* 106 (2009), pp. 16051–6.

d'Errico, F. *et al.*, 'Assessing the accidental versus deliberate colour modification of shell beads: a case study on perforated *Nassarius kraussianus* from Blombos Cave Middle Stone Age levels', *Archaeometry* 57 (2015), pp. 51–76.

d'Errico, F. *et al.*, 'From number sense to number symbols: An archaeological perspective', *Philosophical Transactions B* 373 (2017).

Damerow, P., 'The material culture of calculation: A theoretical framework for a historical epistemology for the concept of number', in U. Gellert and E. Jablonka, eds, *Mathematisation and Demathematisation: Social, Philosophical and Educational Ramifications* (Rotterdam, 2007), pp. 19–56.

Danna, Raffaele, 'Figuring out: the spread of Hindu-Arabic numerals in the European tradition of practical mathematics (13th–16th centuries)', *Nuncius* 36 (2021), pp. 5–48.

Dantzig, Tobias, *Number: The Language of Science* (London, 1930).

Dartnall, T., 'Does the World Leak Into the Mind? Active Externalism, "Internalism" and Epistemology', *Cognitive Science* 29 (2005), pp. 135–43.

Dauben, Joseph W., 'Ancient Chinese mathematics: the (*Jiu Zhang Suan Shu*) *vs* Euclid's *Elements*. Aspects of proof and the linguistic limits of knowledge', *International Journal of Engineering Science* 36 (1998), pp. 1339–59.

Dauben, Joseph W., 'The evolution of mathematics in ancient China: from the newly discovered *Shu* and *Suan Shu Shu* bamboo texts to the *Nine Chapters on the Art of Mathematics'*, *Revista Brasileira de História da Matemática* 19 (2019), pp. 25–78.

Davis, J. T., *Trade Routes and Economic Exchange among the Indians of California* (Berkeley, 1961).

Davis, S. J. M., 'Incised bones from the Mousterian of Kebara cave (Mount Carmel) and the Aurignacian of Ha-Yonim cave (Western Galilee), Israel', *Paléorient* 2 (1974), pp. 181–2.

De Cruz, Helen, 'Numerical cognition and mathematical realism', *Philosophers' Imprint* 16 (2016).

De Cruz, Helen and J. De Smedt, 'The innateness hypothesis and mathematical concepts', *Topoi* 29 (2010), pp. 3–13.

De Cruz, Helen, *et al.*, 'The cognitive basis of arithmetic', in Benedikt Löwe and Thomas Müller, eds, *Philosophy of Mathematics: Sociological Aspects and Mathematical Practice* (London, 2010), pp. 59–106.

de Heinzelin, Jean, *Les Fouilles d'Ishango* (Brussels 1957).

de Heinzelin, Jean, 'Ishango', *Scientific American* 206 (1962), pp. 105–18.

De Odorico, Marco, *The Use of Numbers and Quantifications in the Assyrian Royal Inscriptions* (Helsinki, 1995).

De Vries, Lourens, 'Numerals in Papuan languages of the Greater Awyu family', in Anne Storch and Gerrit J. Dimmendaal, eds, *Number – Constructions and Semantics: Case Studies from Africa, Amazonia, India and Oceania* (Amsterdam, 2014), pp. 329–54.

DeCesare, Richard Patrick, 'Indigenous mathematics of Native North Americans: A sourcebook for educators', unpublished D.Ed. dissertation, Columbia University, 1999.

Delluc, B. and G. Delluc, 'Images de la main dans notre préhistoire', *Les Dossiers d'Archéologie* 178 (1993), pp. 32–45.

Denniss, John, *Figuring It Out: Children's Arithmetical Manuscripts, 1680–1880* (Oxford, 2012).

Depauw, Mark, *The Archive of Teos and Thabis from Early Ptolemaic Thebes: P. Brux. Dem. Inv. E. 8252–8256* (Brussels, 2000).

Deuze, Mark, 'Participation, remediation, bricolage: considering principal components of a digital culture', *The Information Society* 22 (2006), pp. 63–75.

Ditz, H. M. and A. Nieder, 'Neurons selective to the number of visual items in the corvid songbird endbrain', *Proceedings of the National Academy of Sciences* 112 (2015), pp. 7827–32.

Ditz, H. M. and A. Nieder, 'Numerosity representations in crows obey the Weber–Fechner law', *Proceedings of the Royal Society B* 283 (2016).

Dixon, Robert M. W., *The Languages of Australia* (Cambridge, 1980).

Dixon, Robert M. W., *Australian Languages: Their Nature and Development* (Cambridge, 2002).

Dixon, R. B. and A. L. Kroeber, 'Numeral systems of the languages of California', *American Anthropologist* 9 (1907), pp. 663–90.

Djebbar, A., *L'Analyse combinatoire au Maghreb: L'Exemple d'Ibn Mun'im (XIIe–XIIIe s.)* (Orsay, 1985).

Djebbar, A., 'Les pratiques combinatoires au Maghreb à l'époque de Raymond Lulle', *Quaderns de la Mediterrània* 9 (n.d.), pp. 85–91.

Domínguez-Castro, Fernando *et al.*, 'An early weather diary from Iberia (Lisbon, 1631–1632)', *Weather* 70 (2015), pp. 20–4.

Donohue, M., 'Complexities with restricted numeral systems', *Linguistic Typology* 12 (2008), pp. 423–9.

Dorman, Robert L., 'The creation and destruction of the 1890 Federal Census', *The American Archivist* 71 (2008), pp. 350–83.

Dreyer, Günter, 'Recent discoveries at Abydos cemetery U', in Brink, Edwin C. M. van den, ed., *The Nile Delta in Transition: 4th–3rd Millennium BC* (Tel Aviv, 1992), pp. 293–9.

Duhard, J.-P., 'Le calendrier obstétrical de la femme à la corne en bas-relief de Laussel', *Bulletin de la société historioue et archéologique du Périgord* 115 (1988), pp. 23–39.

Duhard, J.-P., 'Etude morphologique de la Femme à la corne en bas-relief de Laussel', *Bulletin de la Société Historique et Archéologique du Périgord* 116 (1989), pp. 257–75.

Duhard, J.-P., 'Le corps féminin et son langage dans l'art paléolithique', *Oxford Journal of Archaeology* 9 (1990), pp. 241–55.

Duhard, J.-P., 'The shape of Pleistocene women', *Antiquity* 65 (1991), pp. 552–61.

Duhard, J.-P., 'The Upper Palaeolithic figures as a reflection of human morphology and social organization', *Antiquity* 67 (1993), pp. 83–91.

Duhard, J.-P., 'L'identité physiologique, un élément d'interprétation des figurations féminines paléolithiques', *Trabajos de Prehistoria* 51 (1994), pp. 39–53.

Duus, Peter, *The Cambridge History of Japan: Volume 6, The Twentieth Century* (Cambridge, 2008).

Dwyer, P. and M. Minnegal, 'Counting systems of the Strickland-Bosavi languages, Papua New Guinea', *Language and Linguistics in Melanesia* 34 (2016), pp. 1–36.

Eikemeier, Peter, 'von Frans Hals bis Vermeer', *Kunstchronik* 37 (1984), pp. 444–8.

Ellerton, Nerida and M. A. (Ken) Clements, *Rewriting the History of School Mathematics in North America 1607–1861: The Central Role of Cyphering Books* (Dordrecht, 2012).

Ellerton, Nerida and M. A. (Ken) Clements, *Abraham Lincoln's Cyphering Book and Ten Other Extraordinary Cyphering Books* (New York, 2014).

Englund, R. K., 'Proto-cuneiform account books and journals', in M. Hudson and C. Wunsch, eds, *Creating Economic Order: Record-keeping, Standardization and the Development of Accounting in the Ancient Near East* (Bethesda, MD, 2004), pp. 23–46.

Epps, P. *et al.*, 'On numeral complexity in hunter-gatherer languages', *Linguistic Typology* 16 (2012), pp. 41–109.

Erskine, Andrew, ed., *A Companion to the Hellenistic World* (Oxford, 2003).

Evans, G. R., 'Difficillima et Ardua: theory and practice in treatises on the abacus 950–1150', *Journal of Medieval History* 3 (1977a), pp. 21–38.

Evans, G. R., 'From abacus to algorism: theory and practice in medieval arithmetic', *The British Journal for the History of Science* 10 (1977b), pp. 114–31.

Evans, G. R., 'Schools and scholars: the study of the abacus in English schools c.980–c.1150', *The English Historical Review* 94 (1979), pp. 71–89.

Evans, M., *Persistence of the Gift: Tongan Tradition in Transnational Context* (Waterloo, 2001).

Evans, N., ed., *The Non-Pama-Nyungan Languages of Northern Australia: Comparative Studies of the Continent's Most Linguistically Complex Region* (Canberra, 2003).

Everett, C. and K. Madora, 'Quantity recognition among speakers of an anumeric language', *Cognitive Science* 36 (2012), pp. 130–41.

Everett, D. L., 'Cultural constraints on grammar and cognition in Pirahã', *Current Anthropology* 46 (2005), pp. 621–46.

Fash, William Leonard and Barbara W. Fash, *Scribes, Warriors, and Kings: The City of Copán and the Ancient Maya* (London, 1991/2001).

Feigenson, L., 'Parallel non-verbal enumeration is constrained by a set-based limit', *Cognition* 107 (2008), pp. 1–18.

Fellmann, Rudolf, 'Römische Rechentafeln aus Bronze', *Antike Welt* 14 (1983), pp. 36–40.

Felt, D. E., *Mechanical Arithmetic, or, The History of the Counting Machine* (Illinois, 1916).

Fienup-Riordan, A., *Eskimo Essays: Yup'ik Lives and how We See Them* (New Brunswick, 1990).

Fienup-Riordan, A. and A. Rearden, *Ellavut / Our Yup'ik World and Weather: Continuity and Change on the Bering Sea Coast* (Seattle, 2012).

Fischer, Martin H., 'A hierarchical view of grounded, embodied and situated numerical cognition', *Cognitive Processing* 38 (2012).

Fleet, Kate *et al.*, eds, *Encyclopaedia of Islam* (3rd edition, brill.com).

Flegg, G., *Numbers, Their History and Meaning* (New York, 1983).

Folkerts, M. ed., *'Boethius' Geometrie II: Ein mathematisches Lehrbuch des Mittelalters* (Wiesbaden, 1970).

Folkerts, M., 'Early texts on Hindu-Arabic calculation', *Science in Context* 14 (2001), pp. 13–38.

Frank, M. and D. Barner, 'Representing exact number visually using mental abacus', *Journal of Experimental Psychology* 141 (2011), pp. 134–49.

Frank, M. C. *et al.*, 'Number as a cognitive technology: Evidence from Pirahã language and cognition', *Cognition* 108 (2008), pp. 819–24.

Freeman, S. E., 'The jetons of the deans of the old faculty of medicine in Paris', *Bulletin of the History of Medicine* 19 (1946), pp. 48–95.

Friedlein, Gottfried, *Die Zahlzeichen und das elementare Rechnen* (Erlangen, 1869).

Froehner, W., 'Le Comput digital', *Annuaire de la Société Française de numismatique et d'archéologie* 8 (1884), pp. 232–8.

Funk, C., 'The Bow and Arrow War Days on the Yukon-Kuskokwim Delta of Alaska', *Ethnohistory* 57 (2010), pp. 523–69.

Gallistel, C. R. 'Finding numbers in the brain', *Philosophical Transactions B* 373 (2017).

Gamble, Lynn H., 'The origin and use of shell bead money in California', *Journal of Anthropological Archaeology* 60 (2020).

Gammage, B., *The sky travellers: Journeys in New Guinea 1938–1939* (Victoria, 1998).

Gandz, Solomon, 'The Origin of the Ghubar Numerals, or the Arabian Abacus and the Articuli', *Isis* 16 (1931), pp. 393–424.

Garcia, M. A. and H. Duday, 'Les empreintes de mains dans l'argile des grottes ornées', *Les Dossiers d'Archéologie* 178, (1993), pp. 56–9.

García-Diez, Marcos *et al.*, 'The chronology of hand stencils in European Palaeolithic rock art: implications of new U-series results from El Castillo Cave (Cantabria, Spain)', *Journal of Anthropological Sciences* 93 (2015), pp. 1–18.

Gelman, R. and B. Butterworth, 'Number and language: How are they related?', *Trends in Cognitive Sciences* 9 (2005), pp. 6–10.

George, Éric, ed., *Digitalization of Society and Socio-Political Issues 1: Digital, Communication, and Culture* (Newark, 2020).

Gere, Charlie, *Digital Culture* (London, 2008).

Giaquinto, M., 'Cognitive access to numbers: the philosophical significance of empirical findings about basic number abilities', *Philosophical Transaction B* 373 (2017).

Gibson, Dominic, 'Gesture's Role in Bridging Symbolic and Nonsymbolic Representations of Number', unpublished Ph.D. dissertation, University of Chicago, 2017.

Gifford, Edward Winslow, *Clear Lake Pomo Society* (California, 1926).

Gifford, E. W., *Tongan Society* (Honolulu, 1929).

Gilligan, I., 'The prehistoric development of clothing: archaeological implications of a thermal model', *Journal of Archaeological Method and Theory* 17 (2010), pp. 15–80.

Gillispie, Charles Coulston *et al.*, eds, *Complete Dictionary of Scientific Biography* (Detroit, 2008).

Ginsburg, Jekuthial, 'New light on our numerals', *Bulletin of the American Mathematical Society* 23 (1917), pp. 366–9.

Glanville, S. R. K., *Catalogue of Demotic Papyri in the British Museum. Volume 1, A Theban Archive of the Reign of Ptolemy I, Soter* (London, 1939).

Goetze, Albrecht, 'Number Idioms in Old Babylonian', *Journal of Near Eastern Studies* 5 (1946), pp. 185–202.

Goldenberg, Gideon, *Semitic Languages: Features, Structures, Relations, Processes* (Oxford, 2013).

Golinski, Jan, '"Exquisite Atmography": theories of the world and experiences of the weather in a diary of 1703', *The British Journal for the History of Science* 34 (2001), pp. 149–71.

Golinski, Jan, *British Weather and the Climate of Enlightenment* (Chicago, 2007).

Gordon, P., 'Numerical Cognition Without Words: Evidence from Amazonia', *Science* 306 (2004), pp. 496–9.

Grant, Lindy, *Blanche of Castile: Queen of France* (New Haven, 2016).

Grayson, Albert Kirk, *et al.*, *Assyrian Rulers of the Third and Second Millennia BC (to 1115 BC)* (Toronto, 1987).

Greenberg, J. H., 'Generalizations about numeral systems', in J. H. Greenberg, ed., *Word Structure: Vol. 3. Universals of Human Language* (Stanford, 1978), pp. 249–95.

Groenen, Marc, 'Les représentations de mains négatives dans les grottes de Gargas et de Tibiran (Hautes-Pyrénées). Approche méthodologique', *Bulletin de la Société royale belge d'Anthropologie et de Préhistoire* 99 (1988), pp. 81–113.

Groenen, Marc, 'Quelques problèmes à propos des mains négatives dans les grottes paléolithiques. Approche épistémologique', *Annales d'Histoire de l'art et d'Archéologie* 12 (1990), pp. 7–29.

Groenen, Marc, 'Images de mains dans la préhistoire. La part de l'œil', *Revue de pensée des arts plastiques* (2011), pp. 56–69.

Groucutt, H. S. *et al.*, 'Rethinking the dispersal of Homo sapiens out of Africa', *Evolutionary Anthropology: Issues, News, and Reviews* 24 (2015), pp. 149–64.

Guitel, Geneviève, *Histoire comparée des numérations écrites* (Paris, 1975).

Gvozdanović, J., ed., *Indo-European Numerals* (Berlin, 1992).

Gvozdanović, J., ed., *Numeral Types and Changes Worldwide* (Berlin, 1999).

Haigh, Thomas and Paul E. Ceruzzi, *A New History of Modern Computing* (Cambridge, MA, 2021).

Halberda, J., *et al.*, 'Multiple spatially overlapping sets can be enumerated in parallel', *Psychological Science* 17 (2006), pp. 572–6.

Hale, K. L., 'Gaps in Grammar and Culture', in M. D. Kinkade *et al.*, eds, *Linguistics and Anthropology: In honour of C. F. Voegelin* (Lisse, 1975), pp. 295–315.

Hammarström, Harald, 'Numerals in the World's Languages: An Update on Status and Interpretation' (2008a), https://lingvist-kredsen.ku.dk/pdf/Oplaeg_Hammerstroem_08.pdf

Hammarström, Harald, 'Complexity in numeral systems with an investigation into pidgins and creoles', in Matti Miestamo, *et al.*, eds, *Language Complexity: Typology, Contact, Change* (Amsterdam, 2008b), pp. 287–304.

Hammarström, Harald, 'Rarities in numeral systems', in J. Wohlgemuth and M. Cysouw, eds, *Rethinking Universals: How Rarities Affect linguistic Theory* (Berlin, 2010), pp. 11–60.

Hannagan, T. *et al.*, 'A randommatrix theory of the number sense', *Philosophical Transactions B* 373 (2017).

Hansen, M. H., *The Athenian Democracy in the Age of Demosthenes* (Oxford, 1991).

Harris, John, 'Facts and Fallacies of Aboriginal Number Systems', *Language and Culture, Work Papers of SIL-AAB, Series B, Vol. 8* (Darwin, 1982), pp. 153–76.

Harris, John, 'Australian Aboriginal and Islander Mathematics', *Australian Aboriginal Studies* 2 (1987), pp. 29–37.

Harrison, Sheldon and Frederick H. Jackson, 'Higher numerals in several Micronesian languages', in B. W. Bender, ed., *Studies in Micronesian Linguistics* (Canberra, 1984), pp. 59–78.

Haselgrove, Colin and Stefan Krmnicek, 'The Archaeology of Money', *Annual Review of Anthropology* 41 (2012), pp. 235–50.

Haskins, C. H., 'The Abacus and the King's Curia', *English Historical Review* 27 (1912), pp. 101–6.

Hatano, G. and K. Osawa, 'Digit memory of grand experts in abacus-derived mental calculation', *Cognition* 15 (1983), pp. 95–110.

Hatano, G. *et al.*, 'Performance of expert abacus operators', *Cognition* 5 (1977), pp. 47–55.

Hatano, G. *et al.*, 'Formation of a mental abacus for computation and its use as a memory device for digits: A developmental study', *Developmental Psychology* 23 (1987), pp. 832–8.

Hatta, T. *et al.*, 'Digit memory of soroban experts: evidence of utilization of mental imagery', *Applied Cognitive Psychology* 3 (1989), pp. 23–33.

Hauser, M. D. *et al.*, 'Spontaneous number representation in semi-free-ranging rhesus monkeys', *Proceedings of the Royal Society B* 267 (2000), pp. 829–33.

Hawke, Jason, 'Number and Numeracy in Early Greek Literature', *Syllecta Classica* 19 (2008), pp. 1–76.

Heide, Lars, 'Punched cards for professional European offices: revisiting the dynamics of information technology diffusion from the United States to Europe, 1889–1918', *History and Technology* 24 (2008), pp. 307–20.

Heide, Lars, *Punched-Card Systems and the Early Information Explosion, 1880–1945* (Baltimore, 2009).

Heizer, Robert F., 'Counterfeiters and shell currency manipulators among California Indians', *Journal of California Anthropology* 2 (1975), pp. 108–10.

Henshilwood, C., 'Fully symbolic sapiens behaviour: Innovation in the Middle Stone Age at Blombos Cave, South Africa', in P. Mellars *et al.*, eds, *Rethinking the Human Revolution: New Behavioural and Biological Perspectives on the Origin and Dispersal of Modern Humans* (Cambridge, 2007), pp. 123–32.

Henshilwood, C. *et al.*, 'An early bone tool industry fron

the Middle Stone Age, Blombos Cave, South Africa: impli-cations for the origins of modern human behaviour, symbolism and language', *Journal of Human Evolution* 41 (2001), pp. 631–78.

Henshilwood, C. *et al.*, 'Engraved ochres from the Middle Stone Age levels at Blombos Cave, South Africa', *Journal of Human Evolution* 57 (2009), pp. 27–47.

Herreman, Alain, 'La mise en texte mathématique: une analyse de l'"Algorisme de Frankenthal"', *Methodos* 1 (2001).

Hilton Turner, J., 'Roman Elementary Mathematics: The Operations', *The Classical Journal* 47 (1951), pp. 63–74, 106–8.

Hishitani, S., 'Imagery experts: How do expert abacus operators process imagery?', *Applied Cognitive Psychology* 4 (1990), pp. 33–46.

Hobart, Michael E. and Zachary Sayre Schiffman, *Information Ages: Literacy, Numeracy, and the Computer Revolution* (Baltimore, 1998).

Hollerith, Herman, 'An Electric Tabulating System', *The School of Mines Quarterly* 10 (1889a), pp. 238–55.

Hollerith, Herman, 'Art of Compiling Statistics', US Patent No. 395,781 (1889b).

Hollerith, Herman, 'The Electrical Tabulating Machine', *Journal of the Royal Statistical Society* 57 (1894), 678–89.

Houston, Stephen D., 'The Archaeology of Communication Technologies', *Annual Review of Anthropology* 33 (2004), pp. 223–50.

Howgego, C., *Ancient History from Coins* (London, 1995).

Huang, Ray, *Taxation and Governmental Finance in Sixteenth-Century Ming China* (Cambridge, 1974).

Hubbard, Edward M. *et al.*, 'The evolution of numerical cognition: from number neurons to linguistic quantifiers', *The Journal of Neuroscience* 28 (2008), pp. 11819–24.

Hudson, J. W., 'Pomo Wampum Makers', *Overland Monthly* 30 (1897), 101–8.

Hudson, M. and C. Wunsch, eds, *Creating Economic Order:*

Record-keeping, Standardization and the Development of Accounting in the Ancient Near East (Bethesda, 2004).

Huehnergard, John, *A Grammar of Akkadian* (3rd edition: Winona Lake, 2011).

Hunt, Terry L. and Ethan E. Cochrane, *The Oxford Handbook of Prehistoric Oceania* (Oxford, 2018).

Hunt, William C., 'The Scope and Method of the Federal Census', *Publications of the American Economic Association* 2 (1899), pp. 466–94.

Hurford, James, *Language and Number* (Oxford, 1987).

Huylebrouck, Dirk, 'The ISShango project', *Journal of Mathematics and the Arts* 2 (2008), pp. 145–52.

Huylebrouck, Dirk, *Africa and Mathematics: From Colonial Findings Back to the Ishango Rods* (Cham, 2019).

Hyde, Daniel C., 'Two systems of non-symbolic numerical cognition', *Frontiers in Human Neuroscience* 5 (2011).

Ifrah, G., *The Universal History of Numbers: From Prehistory to the Invention of the Computer* (London, 1998).

Imhausen, Annette, *Mathematics in Ancient Egypt: A Contextual History* (Princeton, 2016).

Jack, Sybil M., 'A Note on F. P. Barnard, *The casting counter and the counting board: A Chapter in the History of Numismatics and Early Arithmetic*', *Abacus* 3 (1967), pp. 80–2.

Jeffares, B., 'The co-evolution of tools and minds: Cognition and material culture in the hominin lineage', *Phenomenology and the Cognitive Sciences* 9 (2010), pp. 503–20.

Jin, Dal Yong, 'Digital convergence of Korea's webtoons: transmedia storytelling', *Communication Research and Practice* 1 (2015), pp. 193–209.

Jin, Dal Yong, *New Korean Wave: Transnational Cultural Power in the Age of Social Media* (Baltimore, 2016).

Jolles, Carol Zane and Elinor Mikaghaq Oozeva, *Faith, Food, and Family in a Yupik Whaling Community* (Seattle, 2002).

Jones, T. Rupert, 'On Some Implements Bearing Marks Referable to Ownership, Tallies, and Gambling, from the Caves of Dordogne, France', *The Journal of the Anthropological Institute of Great Britain and Ireland* 2 (1873), pp. 362–5.

Jones, Charles W., 'The "Lost" Sirmond Manuscript of Bede's "Computus"', *English Historical Review* 52 (1937), pp. 204–19.

Jones, Charles W., ed., *Bedae Opera de temporibus* (Cambridge, MA, 1943).

Jones, Lindsay, ed., *Encyclopedia of Religion* (2nd edition: Detroit, 2005).

Justeson, J., 'Numerical cognition and the development of "zero" in Mesoamerica', in C. Renfrew and I. Morley, eds., *The Archaeology of Measurement: Comprehending Heaven, Earth and Time in Ancient Societies* (Cambridge, 2010), pp. 43–53.

Kadosh, Roi Cohen and Ann Dowker, eds., *The Oxford Handbook of Numerical Cognition* (Oxford, 2015).

Katz, Victor J., ed., *The Mathematics of Egypt, Mesopotamia, China, India, and Islam: A Sourcebook* (Princeton, 2007).

Katz, Victor J., ed., *Sourcebook in the Mathematics of Medieval Europe and North Africa* (Princeton, 2016).

Keen, I., *Aboriginal Economy and Society: Australia at the Threshold of Colonisation* (Oxford, 2004).

Keyser, Paul, 'The Origin of the Latin Numerals 1 to 1000', *American Journal of Archaeology* 92 (1988), pp. 529–46.

Keyser, Paul, 'Compound Numbers & Numerals in Greek', *Syllecta Classica* 26 (2015), pp. 113–75.

Keyser, Paul, 'Numismatic Evidence for Compound Numbers Written in Greek Alphabetic Numerals', in Nathan T. Elkins and Jane DeRose Evans, eds, *Concordia Disciplinarum: Essays on Ancient Coinage, History, and Archaeology in Honor of William E. Metcale* (New York, 2018) pp. 51–85.

Kim, Ji-Hyeon and Jun Yu, 'Platformizing Webtoons: The Impact on Creative and Digital Labor in South Korea', *Social Media + Society* 5 (2019).

Kim, Soomi, 'A Comparative Study of Korean Abacus Users' Perceptions and Explanations of Use: Including a Perspective on Stigler's Mental Abacus', unpublished Ph.D. dissertation, Columbia University, 2016.

Kim, Youna, ed., *The Routledge Handbook of Korean Culture and Society* (London, 2017).

Kirch, P. K., *On the Road of the Winds: An Archaeological History of the Pacific Islands Before European Contact* (Berkeley, 2000).

Kirsh, D., 'Adapting the environment instead of oneself', *Adaptive Behavior* 4 (1996), pp. 415–52.

Kistermann, F. W. and V. A. Reston, 'The Invention and Development of the Hollerith Punched Card: In Commemoration of the 130th Anniversary of the Birth of Herman Hollerith and for the 100th Anniversary of Large Scale Data Processing', *Annals of the History of Computing* 13 (1991), pp. 245–59.

Klein, Richard G., 'Archaeology and the evolution of human behavior', *Evolutionary Anthropology* 9 (2000), pp. 17–36.

Knecht, Rick and Warren Jones, '"The Old Village": Yup'ik Precontact Archaeology and Community-Based Research at the Nunalleq Site, Quinhagak, Alaska', in *Études Inuit Studies* 43 (2019), pp. 25–52.

Köbel, Jacob, *Ain new geordnet Rechenbiechlin auf den linien mit Rechen pfeningen* (Augsburg, 1514).

Kojima, Takashi, *The Japanese Abacus: Its Use and Theory* (Rutland, VT, 1954).

Kojima, Takashi, *Advanced Abacus* (Rutland, VT, 1963).

Koo, J. H. and G. Bartman, 'The number system in Yupik Eskimo', *Proceedings of the American Philosophical Society* 124 (1980), pp. 48–51.

Kroll, John H., *Athenian Bronze Allotment Plates* (Cambridge, MA, 1972).

Kuhn, S. and M. Stiner, 'Body ornamentation as information technology: towards an understanding of the significance of early beads', in: P. Mellars *et al.*, eds, *Rethinking the Human Revolution: New Behavioural and Biological Perspectives on the Origins and Dispersal of Modern Humans* (Cambridge, 2007), pp. 45–54.

Kuhrt, Amélie, *The Ancient Near East: c.3000–330 B.C.* (London, 1995).

Kunitzsch, Paul, 'The transmission of Hindu-Arabic numerals reconsidered', in Jan P. Hogendijk and Abdelhamid I. Sabra, eds, *The Enterprise of Science in Islam: New Perspectives* (Cambridge, MA, 2003), pp. 3–21.

Kunitzsch, Paul, *Zur Geschichte der 'arabischen' Ziffern* (Munich, 2005).

Kutter, Esther F. *et al.*, 'Single Neurons in the Human Brain Encode Numbers', *Neuron* 100 (2018), pp. 753–61.

Lang, M. L., 'Herodotus and the abacus', *Hesperia* 26 (1957), pp. 271–87.

Lang, M. L., 'The abacus and the calendar', *Hesperia* 33 (1964), pp. 146–67.

Lang, M. L., 'Abaci from the Athenian Agora', *Hesperia* 37 (1968), pp. 241–3.

Lang, M. L., *Ostraka* (Princeton, 1990).

Laycock, D., 'Observations on number systems and semantics', *New Guinea Area Languages and Language Study* 1 (1975), pp. 219–33.

Lean, Glendon, 'Counting systems of Papua New Guinea and Oceania', unpublished DMath dissertation, Papua New Guinea University of Technology, 1992.

Lee, D. S. *et al.*, 'Trans-hemispheric effects of large volcanic eruptions as recorded by an early 19th century diary', *International Journal of Climatology* 30 (2010), pp. 2217–28.

Leeding, Velma, 'Anindilyakwa phonology and morphology', unpublished PhD dissertation, University of Sydney, 1989.

Leibniz, Gottfried Wilhelm, *Confessio Philosophi: Papers Concerning the Problem of Evil, 1671–1678*, edited by R. C. Sleigh *et al.* (New Haven, 2005).

Leitner, Gerhard and Ian G. Malcolm, eds, *The Habitat of Australia's Aboriginal Languages: Past, Present and Future* (Berlin, 2007).

Lemay, Richard, 'The Hispanic origin of our present numeral forms', *Viator* 8 (1977), pp. 435–77.

Lemay, Richard, 'Arabic numerals', in Joseph R. Strayer, ed., *Dictionary of the Middle Ages* (New York, 1982), vol. 1, pp. 382–98.

Lemoine, J., 'Les anciens procedes de calcul sur les doigts en Orient et Occident', *Revue des études islamiques* 6 (1932), pp. 1–56.

Leroi-Gourhan, A., 'Les mains de Gargas. Essai pour une étude d'ensemble', *Bulletin de la Société préhistorique française* 64 (1967), pp. 107–22.

Levy, H. L., 'Catullus, 5, 7–11 and the Abacus', *The American Journal of Philology* 62 (1941), pp. 222–4.

Libbrecht, U., *Chinese Mathematics in the Thirteenth Century: The Shu-shu chiu-chang of Ch'in Chiu-shao* (Cambridge, MA, 1973).

Liddell, Henry George and Robert Scott *et al.*, *A Greek-English Lexicon* (9th edition: Oxford, 1996).

Lie, John, *K-pop: Popular Music, Cultural Amnesia, and Economic Innovation in South Korea* (Berkeley, 2014).

Lieberman, Stephen J., 'Of clay pebbles, hollow clay balls, and writing: a Sumerian view', *American Journal of Archaeology* 84 (1980), pp. 339–58.

Lindemann, Oliver *et al.*, 'Getting a grip on numbers: Numerical magnitude priming in object grasping', *Journal of Experimental Psychology: Human Perception and Performance* 33 (2007), pp. 1400–9.

Lipka, Jerry, 'Culturally Negotiated Schooling: Toward a Yup'ik

Mathematics', *Journal for American Indian Education* 33 (1994), pp. 14–30.

Loeb, Edwin M., *Pomo Folkways* (Berkeley, 1926).

Lorblanchet, M., 'Peindre sur les parois des grottes', *Les Dossiers de l'Archéologie* 46 (1980), pp. 32–9.

Lorrey, Andrew, 'Late Little Ice Age climate variability in New Zealand documented by the Reverend Davis "Dirty Weather" diaries', *Quaternary International* 279–80 (2012), p. 290.

Lorrey, Andrew *et al.*, 'The "dirty weather" diaries of Reverend Richard Davis: insights about early colonial-era meteorology and climate variability for northern New Zealand, 1839–1851', *Climate of the Past* 12 (2016), pp. 553–73.

Loughnane, Robyn and Sebastian Fedden, 'Is Oksapmin Ok? A Study of the Genetic Relationship between Oksapmin and the Ok Languages', *Australian Journal of Linguistics* 31 (2011), pp. 1–42.

Lubar, S., 'Do Not Fold, Spindle or Mutilate: A Social History of the Punch Card', *The Journal of American Culture* 15 (2004), pp. 43–56.

Lu, Yujie, *et al.*, 'Can abacus course eradicate developmental dyscalculia', *Psychology in the Schools* 58 (2021), pp. 235–51.

Lynch, John, *Pacific Languages: An Introduction* (Honolulu, 2018).

Ma, Debin Ma and Richard von Glahn, eds., *The Cambridge Economic History of China, Volume 1: To 1800* (Cambridge, 2022).

Majkić, Ana *et al.*, 'Sequential Incisions on a Cave Bear Bone from the Middle Paleolithic of Pešturina Cave, Serbia', *Journal of Archaeological Method and Theory* 25 (2018), pp. 69–116.

Malafouris, Lambros, 'Grasping the concept of number: How did the sapient mind move beyond approximation?' in C. Renfrew and I. Morley, eds., *The Archaeology of Measurement: Comprehending Heaven, Earth and Time in Ancient Societies* (Cambridge, 2010), pp. 35–42.

Malafouris, Lambros, *How Things Shape the Mind* (Cambridge, MA, 2013).

Mallory, J. P. and D. Q. Adams, *The Oxford Introduction to Proto-Indo-European and the Proto-Indo-European World* (Oxford, 2006).

Mariner, William, *An Account of the Natives of the Tonga Islands* (2nd edition: London, 1818).

Marrou, H. I., 'l'Evangile de vérité et la diffusion du comput digital dans l'antiquité', *Vigiliae Christianae* 12 (1958), pp. 98–103.

Martin, T. C., 'Counting a Nation by Electricity', *Electrical Engineer* (11 November 1891), p. 525.

Martzloff, Jean-Claude, *A History of Chinese Mathematics* (Berlin, 1997).

Masson-MacLean, E. *et al.*, 'Pre-Contact Adaptations to the Little Ice Age in Southwest Alaska: New Evidence from the Nunalleq Site', *Quaternary International* 549 (2020), pp.130–41.

Matang, R., 'Formalising the role of Indigenous counting systems in teaching the formal English arithmetic strategies through local vernaculars: An example from Papua New Guinea', in P. Clarkson *et al.*, eds, *28th Conference of Mathematics Education Research Group of Australasia* (Adelaide, 2005), pp. 505–12.

Matang, R. and K. Owens, K., 'The role of indigenous traditional counting systems in children's development of numerical cognition: Results from a study in Papua New Guinea', *Mathematics Education Research Journal* 26 (2014), pp. 531–53.

Matsuda, Yoshiro, 'Formation of the census system in Japan: 1871–1945 – development of the statistical system in Japan proper and her colonies', *Hitotsubashi Journal of Economics* 21 (1981), pp. 44–68.

May, Sally K. *et al.*, 'Boats on Bark: an Analysis of Groote Eylandt Aboriginal Bark-Paintings featuring Macassan Praus from the 1948 Arnhem Land Expedition, Northern Territory, Australia',

International Journal of Nautical Archaeology 38 (2009), pp. 369–85.

McBrearty, Sally and Alison S. Brooks, 'The revolution that wasn't: a new interpretation of the origin of modern human behavior', *Journal of Human Evolution* 39 (2000), pp. 453–563.

McCauley, Brea *et al.*, 'A Cross-cultural Perspective on Upper Palaeolithic Hand Images with Missing Phalanges', *Journal of Paleolithic Archaeology* 1 (2018), pp. 314–33.

McLellan, James E., *Old Regime France and its Jetons: Pointillist History and Numismatics* (New York, 2020).

McRoberts, Robert, 'Counting at Pularumpi: A Survey of a Traditional Mathematics and Its Implications for Modern Learning', *Aboriginal Child at School* 18 (1990), pp. 19–43.

Mellars, P. *et al.* eds., *Rethinking the Human Revolution: New Behavioural and Biological Perspectives on the Origins and Dispersal of Modern Humans* (Cambridge, 2007).

Meltzer, David J., *First Peoples in a New World: Populating Ice Age America* (2nd edition: Cambridge, 2021).

Menninger, Karl, *Number Words and Number Symbols: A Cultural History of Numbers* (Cambridge, MA, 1969).

Mercier, R., 'Astronomical tables in the twelfth century', in C. Burnett, ed., *Adelard of Bath* (London, 1987), pp. 87–118.

Milbrath, Susan, 'The role of solar observations in developing the preclassic Maya calendar', *Latin American Antiquity* 28 (2017), pp. 88–104.

Miller, Kevin and James Stigler, 'Meanings of Skill: Effects of Abacus Expertise on Number Representation', *Cognition and Instruction* 8 (1991), pp. 29–67.

Miller, P., 'Stimulus number, duration and intensity encoding in randomly connected attractor networks with synaptic depression', *Frontiers in Computational Neuroscience* 7 (2013).

Millet, Nicholas B., 'The Narmer macehead and related objects', *Journal of the American Research Center in Egypt* 27 (1990), pp. 53–9.

Mimica, Jardin, *Intimations of Infinity: The Mythopoeia (Cultural*

Meanings) of the Iqwaye Counting and Number Systems (Oxford, 1988).

Minaud, Gérard, 'Des doigts pour le dire: Le comput digital et ses symboles dans l'iconographie romaine', *Histoire & Mesure* 21 (2006), pp. 3–34.

Mithun, Marianne, *The Languages of Native North America* (Cambridge, 1999).

Miyaoka, O. *et al.*, eds, *The Vanishing Languages of the Pacific Rim* (Oxford, 2007).

Mossolova, A. and R. Knecht, 'Bridging Past and Present: A Study of Pre-Contact Yup'ik Masks from the Nunalleq Site, Alaska', *Arctic Anthropology* 56 (2019), pp. 18–38.

Mote, Frederick W. and Denis C. Twitchett, *The Cambridge History of China: Volume 8, The Ming Dynasty, Part 2, 1368–1644* (Cambridge, 2008).

Needham, Joseph and Ling Wang, *Science and Civilisation in China. Vol. 3, Mathematics and the Sciences of the Heavens and the Earth* (Cambridge, 1959).

Nelson, Edward W., 'The Eskimo about Bering Strait', in *Eighteenth Annual Report of the Bureau of American Ethnology 1896–97* (Washington, DC, 1899).

Netz, Reviel, 'Counter culture: towards a history of Greek numeracy', *History of Science* 40 (2002), pp. 319–32.

Newcomb, H. T., 'The Development of Mechanical Methods of Statistical Tabulation in the United States, with Especial Reference to Population and Mortality Data', in *Transactions of the Fifteenth International Congress on Hygiene and Demography, Washington, September 23–28, 1912: Section IX: Demography* (Washington, DC, 1913), pp. 73–83.

Newsome, Elizabeth A., 'The Trees of Paradise and the Pillars of the World: Vision Quest and Creation in the Stelae Cycle of 18-Rabbit-God K, Copán, Honduras', unpublished PhD dissertation, University of Texas, 1991.

Nieder, Andreas, 'Coding of abstract quantity by number neurons of the primate brain', *Journal of Comparative Physiology A* 199 (2013), pp. 1–16.

Nieder, Andreas, 'The neuronal code for number', *Nature Reviews: Neuroscience* 17 (2016), pp. 366–82.

Nieder, Andreas, 'Evolution of cognitive and neural solutions enabling numerosity judgements: lessons from primates and corvids', *Philosophical Transactions B* 373 (2018).

Nieder, Andreas and S. Dehaene, 'Representation of number in the brain', *Annual Review of Neuroscience* 32 (2009), pp. 185–208.

Nieder, Andreas and E. K. Miller, 'Coding of cognitive magnitude: Compressed scaling of numerical information in the primate prefrontal cortex', *Neuron* 37 (2003), pp. 149–57.

Nieder, Andreas *et al.*, 'A Parieto-Frontal Network for Visual Numerical Information in the Monkey', *Proceedings of the National Academy of Sciences* 101 (2004), pp. 7457–62.

Nieder, Andreas *et al.*, 'Temporal and spatial enumeration processes in the primate parietal cortex', *Science* 313 (2006), pp. 1431–5.

Nissen, H. J. *et al.*, *Archaic Bookkeeping: Early Writing and Techniques of Economic Administration in the Ancient Near East* (Chicago, 1993).

Norberg, Arthur L., 'High-Technology Calculation in the Early 20th Century: Punched Card Machinery in Business and Government', *Technology and Culture* 31 (1990), pp. 753–79.

Normark, Johan, 'Multi-scalar cognitive time: Experiential time, known time, and Maya calendars', *Quaternary International* 405 (2016), pp. 52–60.

Nothaft, C. Philipp E., 'The reception and application of Arabic science in twelfth-century computistics: new evidence from Bavaria', *Journal for the History of Astronomy* 45 (2014), pp. 35–60.

Nothaft, C. Philipp E., 'Medieval Europe's satanic ciphers: on the genesis of a modern myth', *British Journal for the History of Mathematics* 35 (2020), pp. 107–36.

Núñez, R. E., 'Numbers and arithmetic: Neither hardwired nor out there', *Biological Theory* 4 (2009), pp. 68–83.

O'Connor, Stephen *et al.*, 'A weather diary from Donegal, Ireland, 1846–1875', *Weather* 76 (2021–2), pp. 385–91.

O'Mahony, Daniel P., 'Lost But Not Forgotten: The U.S. Census of 1890', *Government Publications Review* 18 (1991), pp. 331–7.

O'Regan, Gerard, *A Brief History of Computing* (London, 2008).

Oppenheimer, Stephen, 'The "express train from Taiwan to Polynesia": on the congruence of proxy lines of evidence', *World Archaeology* 36 (2004), pp. 591–600.

Ormerod, Eleanor Anne, ed., *The Cobham Journals, abstracts and summaries of meteorological and phenological observations* (London, 1880).

Osborne, Robin and Simon Hornblower, *Ritual, Finance, Politics: Athenian Democratic Accounts Presented to David Lewis* (New York, 1994).

Otis, Jessica Marie, 'By the Numbers: Understanding the World in Early Modern England', unpublished PhD dissertation, University of Virginia, 2013.

Overmann, Karenleigh A., 'Material Scaffolds in Numbers and Time', *Cambridge Archaeological Journal* 23 (2013), pp. 19–39.

Overmann, Karenleigh A., 'Finger-Counting in the Upper Palaeolithic', *Rock Art Research* 31 (2014), pp. 63–80.

Overmann, Karenleigh A., 'Beyond Writing: The Development of Literacy in the Ancient Near East', *Cambridge Archaeological Journal* 26 (2016a), pp. 285–303.

Overmann, Karenleigh A., 'Number Concepts Are Constructed Through Material Engagement: A Reply to Sutliff, Read, and Everett', *Current Anthropology* 57 (2016b), pp. 352–6.

Overmann, Karenleigh A., 'The role of materiality in numerical cognition', *Quaternary International* 405 (2016c), pp. 42–51.

Overmann, Karenleigh A., 'Thinking Materially: Cognition as Extended and Enacted', *Journal of Cognition and Culture* 17 (2017a), pp. 354–73.

Overmann, Karenleigh A., 'Materiality and Numerical Cognition: A Material Engagement Theory Perspective', in T. Wynn and F. L. Coolidge, eds, *Cognitive Models in Palaeolithic Archaeology* (Oxford, 2017b), pp. 89–112.

Overmann, Karenleigh A. *The Material Origin of Numbers: Insights from the Archaeology of the Ancient Near East* (Piscataway, 2019).

Overmann, Karenleigh A. *et al.*, 'The Prehistory of Number Concept', *Behavioral and Brain Sciences* 34 (2011), pp. 142–4.

Owens, Kay, 'The work of Glendon Lean on the counting systems of Papua New Guinea and Oceania', *Mathematics Education Research Journal* 13 (2001), pp. 47–71.

Owens, Kay *et al.*, *History of Number: Evidence from Papua New Guinea and Oceania* (Cham, 2018).

Palmer, E. H., 'Explanation of a difficult passage in Firdausi', *Journal of Philology* 2 (1869), p. 247.

Parker, W., 'A study of shell bead context, distribution, and use within Northern California', unpublished MA dissertation, California State University, 2010.

Pasfield-Neofitou, Sarah E. *et al.*, *Manga Vision: Cultural and Communicative Perspectives* (Clayton, 2016).

Perey, Arnold, 'Oksapmin society and world view', unpublished PhD dissertation, Columbia University, 1973.

Periton, Cheryl, 'The medieval counting table revisited: a brief introduction and description of its use during the early modern period', *Bulletin of the British Society for the History of Mathematics* 30 (2015), pp. 35–49.

Pestman, P. W., *Familiearchieven uit het land van Pharao: een bundel artikelen samengesteld naar aanleiding van een serie lezingen van het Papyrologisch Instituut van de Rijksuniversiteit van Leiden in het voorjaar van 1986* (Zutphen, 1989).

Petersen, G., 'Indigenous island empires: Yap and Tonga considered', *Journal of Pacific History* 35 (2000), pp. 5–27.

Pettitt, P. B. *et al.*, 'Are hand stencils in European cave art older than we think? An evaluation of the existing data and their potential implications', in Primitiva Bueno Ramírez and Paul G. Bahn, eds, *Prehistoric Art as Prehistoric Culture: Studies in Honour of Professor Rodrigo de Balbin-Behrmann* (Oxford, 2015), pp. 31–43.

Pharo, Lars Kirkhusmo, *The Ritual Practice of Time: Philosophy and Sociopolitics of Mesoamerican Calendars* (Leiden, 2014).

Piazza, M. *et al.*, 'Tuning curves for approximate numerosity in the human intraparietal sulcus', Neuron 44 (2004), pp. 547– 55.

Pica, P. *et al.*, 'Exact and approximate arithmetic in an Amazonian indigene group', *Science* 306 (2004), pp. 499–503.

Pineda De Carías, María C. *et al.*, 'Stela D: A sundial at Copan, Honduras', *Ancient Mesoamerica* 28 (2017), pp. 543–57.

Plank, F. 'Senary summary so far', *Linguistic Typology* 13 (2009), pp. 337–45.

Pletser, Vladimir and Dirk Huylebrouck, 'The Ishango Artifact: The Missing Base 12 Link', *Forma* 14 (1999), pp. 339–46.

Plofker, K., *Mathematics in India* (Princeton, 2009).

Plofker, K., 'Spoken Text and Written Symbol: The Use of Layout and Notation in Sanskrit Scientific Manuscripts', *Digital Proceedings of the Lawrence J. Schoenberg Symposium on Manuscript Studies in the Digital Age* 1 (2009b), article 3.

Poovey, Mary, *A History of the Modern Fact: Problems of Knowledge in the Sciences of Wealth and Society* (Chicago, 1998).

Porter, Robert P., 'The Eleventh Census', *Publications of the American Statistical Association* 2 (1891), pp. 330, 339.

Porter, Robert P., 'The Eleventh United States Census', *Journal of the Royal Statistical Society* 57 (1894), pp. 643–77.

Porter, Theodore M., *Trust in Numbers: The Pursuit of Objectivity in Science and Public Life* (Princeton, 1995).

Powers, Stephen, *Tribes of California* (Washington, DC, 1877).

Price, M., 'Thoughts on the beginnings of coinage', in C. Brooke et al., eds, *Studies in Numismatic Method presented to Philip Grierson* (Cambridge, 1983), pp. 1–10.

Pritchett, W. Kendrick, 'Gaming tables and IG 12 324', *Hesperia* 34 (1965), pp. 131–47.

Pritchett, W. Kendrick, '"Five lines" and IG 12 324', *California Studies in Classical Antiquity* 1 (1968), pp. 187–215.

Pullan, J. M., *The History of the Abacus* (London, 1968).

Rathbone, Perry T., 'The Housekeeper by Nicolaes Maes', *Bulletin of the City Art Museum of St Louis* 36 (1951), pp. 42–5.

Reese, D. S., 'On the incised cattle scapulae from the East Mediterranean and Near East', *Bonner Zoologische Beitrage* 50 (2002), pp. 183–98.

Reid-Green, Keith S., 'The History of Census Tabulation', *Scientific American* 260 (1989), pp. 98–103.

Rhee, Younghoon, 'A comparative historical study of the census registers of early Choson Korea and Ming China', *International Journal of Asian Studies* 2 (2005), pp. 25–55.

Richardson, Leon J., 'Digital Reckoning among the Ancients', *American Mathematical Monthly* 23 (1916), pp. 7–12.

Robinson, J. E., 'Neither Use nor Ornament: A Consideration of the Evidence for the Existence of a System of Communication and Notation in the European Upper Palaeolithic', unpublished Ph. D. dissertation, University of Durham, 1993.

Robinson, William Walker, 'The early works of Nicolaes Maes, 1653 to 1661', unpublished Ph.D. dissertation, Harvard University, 1996.

Robson, Eleanor, *Mathematics in Ancient Iraq: A Social History* (Princeton, 2008).

Rödiger, Emil, 'Ueber die im Orient gebräuliche Fingersprache für den Ausdruck der Zahlen', *Jahresbericht der deutschen morgen-ländischen Gesellschaft 1845* (1846), pp. 111–129.

Rouillon, A., 'Au Gravettien, dans la grotte Cosquer (Marseille, Bouches-du-Rhône), l'Homme a-t-il compté sur ses doigts?', *Anthropologie* 110 (2006), pp. 500–509.

Rowlands, M., *The new science of the mind: From extended mind to embodied phenomenology* (Cambridge, MA, 2010).

Rowlandson, Jane, *Women and Society in Greek and Roman Egypt: a Sourcebook* (Cambridge, 1998).

Rubincam, C., 'Casualty Figures in the Battle Descriptions of Thucydides', *Transactions of the American Philological Association* 121 (1991), pp. 181–198.

Rubincam, C., 'The topography of Pylos and Sphakteria and Thucydides' measurements of distance', *The Journal of Hellenic studies* 121 (2001), pp.77–90.

Rubincam, C., 'Numbers in Greek Poetry and Historiography: Quantifying Fehling', *Classical Quarterly* 53 (2003), pp. 448–63.

Ruggles, Steven and Diana L. Magnuson, 'Census Technology, Politics, and Institutional Change, 1790–2020', *The Journal of American history* 107 (2020), pp. 19–51.

Rutherford, N., ed., *Friendly Islands: A history of Tonga* (Melbourne, 1977).

Saidan, Ahmad Salim, *The Arithmetic of Al-Uqlidisi* (Dordrecht 1978).

Sanderson, M. G., 'Daily weather in Dublin 1716–1734: the diary of Isaac Butler', 73 (2018), pp. 179–182.

Sarmant, Thierry and François Ploton-Nicollet, *Jetons des institutions centrales de l'ancien régime* (Paris, 2010–).

Sartoretto S. *et al.*, 'Quand la Grotte Cosquer a-t-elle été fermée par la montée des eaux?', *Revue Méditerranée* 82 (1995), pp. 21–24.

Saxe, Geoffrey B., 'Body Parts as Numerals: A Developmental Analysis of Numeration among the Oksapmin in Papa New Guinea', *Child Development* 52 (1981), pp. 306–316.

Saxe, Geoffrey B., 'Developing Forms of Arithmetic Operations Among the Oksapmin of Papua New Guinea', *Developmental Psychology* 18 (1982), pp. 583–594.

Saxe, Geoffrey B., 'Cognition, development, and cultural practices', in E. Turiel, ed., *Development and cultural change: Reciprocal processes* (San Francisco, 1999), pp. 19–35.

Saxe, Geoffrey B. and Indigo Esmonde, 'Studying cognition in flux: a historical treatment of *fu* in the shifting structure of Oksapmin mathematics', *Mind, Culture and Activity* 12 (2005), pp. 171–225.

Saxe, Geoffrey B., *Cultural development of mathematical ideas: Papua New Guinea Studies* (Cambridge, 2012).

Sayers, Barbara, 'Aboriginal mathematical concepts: a cultural and linguistic explanation for some of the problems', *Language and Culture: Work papers of SIL-AAB* 8 (1982), pp. 183–200.

Schaps, David M., *The Invention of Coinage and the Monetization of Ancient Greece* (Ann Arbor, 2004).

Schärlig, Alain, 'Les deux types d'abaques des anciens grecs et leurs jetons quinaires', *Archives des Sciences* 54 (2001a), pp. 69–75.

Schärlig, Alain, *Compter avec des cailloux: Le calcul élémentaire sur L'abaque chez les anciens Grecs* (Lausanne, 2001).

Schärlig, Alain, *Compter avec des jetons: Tables à calculer et tables de compte du Moyen Age à la Révolution* (Lausanne, 2003).

Schärlig, Alain, *Compter du bout des doigts - Cailloux, jetons et bouliers, de Périclès à nos jours* (Lausanne, 2006).

Schärlig, Alain, *Du zéro à la virgule: les chiffres arabes à la conquête de l'Europe: 1143–1585* (Lausanne, 2010).

Scheidel, W., 'Finances, Figures and Fiction', *Classical Quarterly* 46 (1996), pp. 222–38.

Schele, Linda and Peter Mathews, *The Code of Kings. The Language of Seven Sacred Maya Temples and Tombs* (New York, 1998).

Schmandt-Besserat, D., *Before writing: From counting to cuneiform* (Austin, TX, 1992).

Schöll, F., *Geschichte der griechischen Litteratur III* (Berlin, 1830).

Schor, Paul, *Counting Americans: How the US Census Classified the Nation* (New York, 2017).

Scriba, Christoph J. and M. E. Dormer Ellis, *The concept of number. A chapter in the history of mathematics* (Mannheim, 1969).

Seth, Michael J., *A Concise History of Modern Korea: From the Late Nineteenth Century to the Present* (3rd edition: Lanham, 2020).

Shapiro, L., *Embodied Cognition* (2nd edition: London, Routledge, 2019).

Sharer, Robert J. and Loa P. Traxler, *The Ancient Maya* (6th edition: Stanford, 2006).

Shaw, Robert D., 'An Archaeology of the Central Yupik: A Regional Overview for the Yukon-Kuskokwim Delta, Northern Bristol Bay, and Nunivak Island', *Arctic Anthropology* 35 (1998), pp. 234–46.

SHCA (Subcommittee of House Committee on Appropriations), Sundry Civil Appropriation Bill for 1903.

Shim, Aegyung, *et al.*, 'Cultural intermediation and the basis of trust among webtoon and webnovel communities', *Information, Communication & Society* 23 (2020), pp. 833–48.

Silva, Luis Pedro, 'Climate and crops in northwest Portugal (1798–1830): A glimpse into the past by the light of two Benedictine diaries', *Historia Agraria* 82 (2020), pp. 99–139.

Sing, Robert, *et al.*, eds., *Numbers and Numeracy in the Greek Polis* (Leiden, 2021).

Sloan, Anna C., 'Gender, Identity, and Belonging: A Community-Based Social Archaeology of the Nunalleq Site in Quinhagak, Alaska', unpublished PhD dissertation, University of Oregon, 2021.

Smith, David Eugene, *Computing Jetons* (New York, 1921).

Smith, David Eugene and Louis Charles Karpinski, *The Hindu-Arabic Numerals* (Boston, 1911).

Staveley, E. S., *Greek and Roman Voting and Elections* (London, 1972).

Stearns, Robert E. C., 'Shell-money', *The American Naturalist* 3 (1869), pp. 1–5.

Stearns, Robert E. C., 'Aboriginal Shell Money', 11 (1877), pp. 344–50.

Stevens, Wesley, 'Hrabanus Maurus on reckoning', unpublished PhD dissertation, Emory University, 1968.

Stewart, Omer C., *Notes on Pomo Ethnogeography* (Berkeley, 1943).

Stigler, James W., 'Abacus skill in Chinese children: Imagery in mental calculation', unpublished PhD dissertation, University of Michigan, 1982.

Stigler, James W., '"Mental abacus": the effect of abacus training on Chinese children's mental calculation', *Cognitive Psychology* 16 (1984), pp. 145–76.

Stigler, James W. *et al.*, 'Motor and mental abacus skill: A preliminary look at an expert', *Quarterly Newsletter of the Laboratory of Comparative Human Cognition* 4 (1982) pp. 12–14.

Stigler, James W. *et al.*, 'Consequences of skill: The case of abacus training in Taiwan', *American Journal of Education* 94 (1986), pp. 447–79.

Stoianov, I. and M. Zorzi, 'Emergence of a "visual number sense" in hierarchical generative models', *Nature. Neuroscience* 15 (2012), pp. 194–6.

Stokes, Judith, 'A description of the mathematical concepts of Groote Eylandt Aborigines', *Language and Culture, Work Papers of SIL-AAB, Series B, Vol. 8* (Darwin, 1982), pp. 33–152.

Suchtelen, Ariane van *et al.*, *Nicolaes Maes* (The Hague, 2019).

Sugden, Keith F., 'A history of the abacus', *Accounting Historians Journal* 8 (1981), pp. 1–22.

Szemerency, O., *Studies in the Indo-European System of Numerals* (Heidelberg, 1960).

Tannery, Paul, 'Notice sur les deux lettres arithmétiques de Nicolas Rhabdas', *Notices et Extráits des Manuscrits de la Bibliothèque Nationale* 32 (1886), pp. 121–252.

Terrell, John Edward, *Prehistory of the Pacific Islands* (Cambridge, 1986).

Thornton, Ken *et al.*, 'Algernon Henry Belfield and the Eversleigh weather diaries, 1877–1922', *Journal of Australian Colonial History* 20 (2018), pp. 139–54.

Tod, Marcus Niebuhr, 'The Greek numeral notation', *Annual of the British School at Athens* 18 (1911–12), pp. 98–132.

Tod, Marcus Niebuhr, 'The Greek numeral systems', *Journal of Hellenic Studies* 33 (1913), pp. 27–34.

Tod, Marcus Niebuhr, 'Further notes on the Greek acrophonic numerals', *Annual of the British School at Athens* 28 (1926–27), pp. 141–157.

Tod, Marcus Niebuhr, 'The Greek acrophonic numerals', *Annual of the British School at Athens* 37 (1936–37), pp. 236–57.

Tod, Marcus Niebuhr, 'The alphabetic numeral system in Attica', *Annual of the British School at Athens* 45 (1950), pp. 126–39.

Tod, Marcus Niebuhr, 'Letter-labels in Greek inscriptions', *Annual of the British School at Athens* 49 (1954), pp. 1–8.

Tod, Marcus Niebuhr, *Ancient Greek Numerical Systems* (Chicago, 1979).

Treacy, Kaye *et al.*, 'Starting points and pathways in Aboriginal students' learning of number: recognising different world views', *Mathematics Education Research Journal* 27 (2015), pp. 263–81.

Trubitt, Mary Beth D., 'The production and exchange of marine shell prestige goods', *Journal of Archaeological Research* 11 (2003), pp. 243–77.

Truesdell, Leon E., *The Development of Punched Card Tabulation in the Bureau of the Census 1890–1940* (Washington, DC, 1965).

Turckheim-Pey, Sylvie de, *Le jeton au Moyen Âge: vers 1250–1498* (Paris, 1997).

Turner, David H., *Tradition and Transformation: A Study of the Groote Eylandt Area Aborigines of Northern Australia* (Canberra, 1974).

Turner, Frederick Jackson, *The Frontier in American History* (Tucson, 1986).

Turner, J. Hilton, 'Roman Elementary Mathematics: The Operations', *The Classical Journal* 47 (1951), pp. 63–74, 106–8.

Uller, C. *et al.*, 'Spontaneous representation of number in cotton-top tamarins (*Saguinus oedipus*)', *Journal of Comparative Psychology* 115 (2001), pp. 248–57.

USCB (US Census Bureau), *100 Years of Data Processing: The Punchcard Century* (Washington, DC, 1991).

USCB (US Census Bureau), *Measuring America: The Decennial Censuses from 1790 to 2000* (Washington, DC, 2002).

Valério, Miguel and Silvia Ferrara, 'Numeracy at the dawn of writing: Mesopotamia and beyond', *Historia Mathematica* 59 (2020), pp. 35–53.

van Berkel, T. A., 'Voiced Mathematics: Orality and Numeracy', in Niall Slater, ed., *Voice and Voices in Antiquity: Orality and Literacy in the Ancient World, Vol. 11* (Leiden, 2017), pp. 321–50.

Van De Mieroop, Marc, *A History of the Ancient Near East, ca. 3000–323 BC* (Oxford, 2004).

van Egmond, Marie-Elaine and Brett Baker, 'The genetic position of Anindilyakwa', 40 (2020), pp. 492–527.

Vanhaeren, M. *et al.*, 'Middle Paleolithic shell beads in Israel and Algeria', *Science* 312 (2006), pp. 1785–8.

Vanhaeren, M. *et al.*, 'Thinking strings: Additional evidence for personal ornament use in the Middle Stone Age at Blombos Cave, South Africa', *Journal of Human Evolution* 64 (2013), pp. 500–17.

Vayda, Andrew P., 'Pomo Trade Feasts', in G. Dalton, ed., *Tribal and Peasant Economies: Readings in Economic Anthropology* (Garden City, NY, 1967).

Verdan, Samuel, 'Counting on Pots? Reflections on Numerical Notations in Early Iron Age Greece', in J. Strauss Clay *et al.*, eds, *Graphê in Late Geometric and Protoarchaic Methone, Macedonia (ca. 700 BCE)* (Berlin, 2017), pp. 105–22.

Verguts, T. and W. Fias, 'Representation of number in animals and humans: a neural model', *Journal of Cognitive Neuroscience* 16 (2004), pp. 1493–1504.

Volkov, Alexei, 'Large numbers and counting rods', in Alexei Volkov, ed., *Sous les nombres, le monde* (Paris, 1994), pp. 71–91.

Volkov, Alexei, 'Le bacchette', in Karine Chemla, ed., *Enciclopedia Italiana, vol. II: La Scienza in Cine, India, Americhe, section I: Storia della Scienza* (Rome, 2002), pp. 125–33.

Volkov, Alexei, 'Chinese Counting Rods: Their History, Arithmetic Operations, and Didactic Repercussions', in Alexei Volkov and Viktor Freiman, eds, *Computations and Computing Devices in Mathematics Education Before the Advent of Electronic Calculators* (Cham, 2018), pp. 137–88.

von Fichtenau, H., 'Wolfger von Prüfening', *Mitteilungen des österreichischen Instituts für Geschichtsforschung* 51 (1937), pp. 313–57.

von Reden, S., 'Money, Law and Exchange: Coinage in the Greek Polis', *Journal of Hellenic Studies* 127 (1997), pp. 154–76.

Walker, C. B. F., *Cuneiform* (London, 1987).

Walker, Francis A., 'The Eleventh Census of the United States', *Quarterly Journal of Economics* 2 (1888), pp. 135–61.

Wallis, Faith, *The Reckoning of Time* (Liverpool, 1999).

Walsh, G. L., 'Mutilated hands or signal stencils? A consideration of irregular hand stencils from central Queensland', *Australian Archaeology* 9 (1979), pp. 33–41.

Wang, Chunjie, 'A Review of the Effects of Abacus Training on

Cognitive Functions and Neural Systems in Humans', *Frontiers in Neuroscience* 14 (2020).

Wardhaugh, Benjamin, *Poor Robin's Prophecies: A Curious Almanac, and the Everyday Mathematics of Georgian Britain* (Oxford, 2012).

Watkins, Jane Iandola *et al.*, *Masters of Seventeenth-Century Dutch Genre Painting* (Philadelphia, 1984).

Weddell, Moritz, 'Actio – loquela digitorum – computatio: Zur Frage nach dem numerus zwischen Ordnungsangeboten, Gebrauchsformen und Erfahrungsmodalitäten', in Moritz Weddell, *Was zählt: Ordnungsangebote, Gebrauchsformen und Erfahrungsmodalitäten des »numerus« im Mittelalter* (Cologne, 2012), pp. 15–63.

Wedell, Moritz, 'Numbers', in Albrecht Classen, ed., *Handbook of Medieval Culture: Fundamental Aspects and Conditions of the European Middle Ages* (Berlin, 2015), pp. 1205–60.

Welborn, Mary Catherine, 'Ghubar numerals' [letter], *Isis* 37 (1932), pp. 260–3.

Wiese, Heike, *Numbers, Language, and the Human Mind* (Cambridge, 2003).

Wiese, Heike, 'The co-evolution of number concepts and counting words', *Lingua* 117 (2007), pp. 758–72.

Wilcox, Walter F., 'The Development of the American Census Office since 1890', *Political Science Quarterly* 29 (1914), pp. 438–59.

Williams, B. P. and R. S. Williams, 'Finger numbers in the Greco-Roman world and the Early Middle Ages', *Isis* 86 (1995), pp. 587–608.

Williams, Maria Sháa Tláa, *The Alaska Native Reader: History, Culture, Politics* (Durham, NC, 2009).

Wilson, N. G., 'Miscellanea Palaeographica', *Greek, Roman and Byzantine Studies* 22 (1981), pp. 395–404.

Woods, C. and L. Feliu, 'The abacus in Mesopotamia: considerations from a comparative perspective', in L. Feliu *et al.*, eds, *The*

First Ninety Years: A Sumerian Celebration in Honor of Miguel Civil (Berlin, 2017), pp. 416–78.

Worsley, Peter, 'The Changing Social Structure of the Wanindiljaugwa', unpublished PhD dissertation, Australian National University, 1954.

Wright, Carroll D., 'How a Census Is Taken', *North American Review* 148 (1889), pp. 727–37.

Wright, Carroll D., *The History and Growth of the United States Census, prepared for the Senate Committee on the Census* (Washington, DC, 1900).

Wyatt, W. F., 'Fractional Quantities on the Abacus', *The Classical Journal* 59 (1964), pp. 268–71.

Wyllys Andrews, Edward and William Leonard Fash, *Copán: The History of an Ancient Maya Kingdom* (Santa Fe, 2005).

Wynn, Thomas *et al.*, 'The archaeology of number concept and its implications for the evolution of language', in R. Botha and M. Everaert, eds., *The Evolutionary Emergence of Human Language: Evidence and Inference* (Oxford, 2013), pp. 118–38.

Wynn, Thomas *et al.*, 'Bootstrapping Ordinal Thinking', in T. Wynn and F. L. Coolidge, eds, *Cognitive Models in Palaeolithic Archaeology* (Oxford, 2017), pp. 197–213.

Yecies, Brian, *et al.*, 'Global transcreators and the extension of the Korean webtoon IP-engine', *Media, Culture and Society* 42 (2020), pp. 40–57.

Yiwen, Zhu, 'On Oin Jiushao's writing system', *Archive for History of Exact Sciences* 74 (2020), pp. 345–79.

Zhang, Jiajie and Hongbin Wang, 'The effect of external representations on numeric tasks', *The Quarterly Journal of Experimental Psychology A: Human Experimental Psychology* 58 (2005), pp. 817–38.

Zhang, Wenxian, 'The Yellow Register Archives of Imperial Ming China', *Libraries & the Cultural Record* 43 (2008), pp. 148–75.

Zhang, Xue-Zhen, 'Precipitation variations in Beijing during

1860–1897 AD revealed by daily weather records from the Weng Tong-He Diary', *International Journal of Climatology* 33 (2013), pp. 568–76.

Zhou, K. and C. Bowern, 'Quantifying uncertainty in the phylogenetics of Australian numeral systems', *Proceedings of the Royal Society B: Biological Sciences* 282 (2015).

Zorzi, M. and A. Testolin, 'An emergentist perspective on the origin of number sense', *Philosophical Transactions B* 373 (2017).

Index

Page references in *italics* indicate images.